HOWELL'S STORM

HOWELL'S
STORM

NEW YORK CITY'S
OFFICIAL RAINMAKER
AND THE 1950 DROUGHT

JIM LEEKE

CHICAGO
REVIEW
PRESS

Published by Chicago Review Press Incorporated
814 North Franklin Street
Chicago, Illinois 60610
ISBN 978-0-912777-95-5

Library of Congress Cataloging-in-Publication Data
Names: Leeke, Jim, 1949– author.
Title: Howell's storm : New York City's official rainmaker and the 1950
    drought / Jim Leeke.
Description: Chicago, Illinois : Chicago Review Press Incorporated, [2019] |
    Includes bibliographical references and index.
Identifiers: LCCN 2018050844 (print) | LCCN 2018057703 (ebook) | ISBN
    9780912777962 (PDF edition) | ISBN 9780912777979 (EPUB edition) | ISBN
    9780912777986 (Kindle edition) | ISBN 9780912777955 (cloth edition)
Subjects: LCSH: Rain-making—New York (State)—New York. | Howell,
    Wallace E., –1999. | Precipitation (Meteorology)—Modification—
    New York (State)—New York. | Droughts—New York (State)—New York—
    History—20th century.
Classification: LCC QC928.6 (ebook) | LCC QC928.6 .L44 2019 (print) |
    DDC 551.68/76092 [B] —dc23
LC record available at https://lccn.loc.gov/2018050844

Typesetting: Nord Compo

Printed in the United States of America
5 4 3 2 1

In memory of my parents, Jim and Betty,
who greeted me nine hours before 1950 began.

"Frown-lines written in cloud. The fish-scale gleam of rain."
—Harry Bingham, *This Thing of Darkness*

"We long for rain especially when we've gone without."
—Cynthia Barnett, *Rain: A Natural and Cultural History*

"We don't believe in rainmakers."
"What do you believe in, mistah? Dyin' cattle?"
—Burt Lancaster as Bill Starbuck, *The Rainmaker* (1956)

Contents

Introduction

The *New York Times* noted the elderly meteorologist's passing nearly a month after his death in San Diego in 1999: WALLACE E. HOWELL, 84, DIES; FAMED RAINMAKER IN DROUGHT. The obituary explained that his "celebrated rain making efforts during a water shortage 50 years ago were admired by parched New Yorkers and detested by drenched Catskill farmers." An old photo showed him standing outside city hall amid a few inches of snow, bareheaded, collar flipped up, hands thrust into his overcoat pockets. His expression was enigmatic; he might have been suppressing a smile. How many *Times* readers remembered Dr. Howell or the drought of 1949–1950? Only a few, surely, although one reader in the hinterlands beyond the Hudson River clipped the obituary and filed it away. Any good story offers opportunity.

If one can pinpoint the beginning and end of a gradual natural phenomenon, one might say the New York City drought lasted eighteen months minus a day. During that period, authorities tried to alleviate a growing water crisis by modifying the weather. It was, after all, the middle of the twentieth century, the American Century, when New Yorkers in particular saw themselves as a people "to whom nothing is impossible." Eight million dusty, dry, and occasionally unshaven residents looked to science for solutions. City hall handed funds and resources to Howell, a modest young World War II veteran from Harvard University.

He believed, but never promised, that he could coax additional rainfall from storms as they rumbled over the city's watershed in the Catskill Mountains.

Wallace Howell began his work amid fears and expectations, tempests and controversy, science and roughhouse politics. In the end, the reservoirs filled; the emergency passed. Howell was both cursed and praised, and then sent on his way to be largely forgotten. Therein lies a tale.

1

Drought

New York City began 1949 both wet and dry. Recent rain and snow had helped to ease a long dry spell that took hold the previous year. But the Kensico Reservoir, thirty miles northeast of Times Square, from which the city drew much of its water, was at its lowest January level ever. "Colonial roads, stone walls and remnants of buildings that lay in the valley before Kensico Dam formed the reservoir came to view again as they did last autumn," the *New York Times* reported. The water level had fallen nearly eleven feet below the spillway, "and a parched perimeter of rocks made it seem even lower."

Winter gave way to a dry spring across much of the state. On Thursday, May 26, as New Yorkers looked forward to a cool but sunny Memorial Day weekend, measurable rain fell on the city. Except for 0.01 inches that would moisten the city before dawn on June 19, no meaningful precipitation fell for another forty-one days. New York's Central Park historically averages nearly 50 inches of rain per year. This year, it would receive 36.25 inches, a shortfall of more than one-quarter.

City parks, rooftops, and pavements began baking in the sunshine. By mid-June, most of New York State was scorching. Temperatures in many places were already edging toward 100 degrees Fahrenheit. Before it ended, 1949 would become the "warmest year in the recorded meteorological history of New York" to that point, with the mean temperature nearly 4.5 degrees above normal. A dozen forest fires burned in the Adirondacks, with little likelihood of rainstorms to douse them. "At Ithaca, the State College of Agriculture warned that a 'first class drought-emergency' was in prospect over the state unless rains came soon." An editorial in the *Times* could have been written by dozens of small-town editors upstate: "Every fisherman and camper, every motorist, every smoker who strolls along the highway should be on guard against the careless spark and the hot ash. Watchfulness is the duty of every citizen while he waits for the reviving rain."

Humidity topped 90 percent in the city but brought no storms. Two million people flocked to the beaches one rainless but "disgustingly damp" Sunday. Even the most jaded New Yorkers began to take notice, perhaps remembering John Steinbeck's Dust Bowl novel *The Grapes of Wrath*, published ten years earlier, in which stoic men "stood by their fences and looked at the ruined corn, drying fast now, only a little green showing through the film of dust." Noted the *Brooklyn Eagle*, "Drought ordinarily is something that happens to other people. Farmers, maybe, and poor folk out in Oklahoma. . . . But when almost five weeks go by with scarcely a drop of rain, something begins to disturb even the city man."

By the Fourth of July, the New York State canal system had restricted boat movements in upstate canals because of low water. Barge traffic on important industrial canals faced delays. Water levels in upstate reservoirs steadily declined, while anxieties steadily rose. Rain finally dampened the five boroughs again on the morning of July 6, dropping temperatures 8 degrees in two hours. "A sudden, short-lived thunder shower descended on the New York metropolitan area, bringing brief respite to New Yorkers suffering from seven weeks of heat and drought," the *Eagle* reported. A bolt of lightning struck a house on Shore Parkway, causing a woman to faint in her kitchen. The storm was otherwise unremarkable, delivering less than 0.25 inches of rain. It was merely a respite, a reminder of blessedly normal summers. "The first few drops make the dust jump. It seems impossible that drops of water could raise a dust cloud in a garden, but they do," the *New York Times* observed.

People began comparing 1949 to other great drought years. "An archivist, writing of the water famine of 1910, said it was a condition 'that was without parallel in the City's history and which it is certain will never happen again.'" Conditions were equally bad all across the Northeast. An official in Middlesex County, New Jersey, began talking about a new approach to drought relief, turning to an emerging field that scientists were beginning to call *experimental meteorology*. "He said he had prepared a plan to seed clouds over Middlesex County with dry ice, a rain-making strategy developed by scientists." Seeding the clouds opposite southern Staten Island likely wouldn't have accomplished much even if it had worked, since most of the rainfall would have run off into Raritan Bay. But the idea perhaps got other politicians thinking.

In late September, the US Department of Agriculture designated twelve New York counties as disaster areas because of drought damage during the summer. A month later, New York City was "teetering on the brink of water rationing—only .1 of an inch of rain has fallen so far this month against a normal 3.53 inches." The Kensico Reservoir was less than half full, "the lowest level since it was added to the city water system in 1916. The shore line has receded 175 yards from the dam, leaving a barren expanse of cracked, sandy soil. Other reservoirs look like receding oases in the desert."

All the while, water consumption in New York City climbed. Officials said residents needed to reduce consumption by at least 200 million gallons a day. Some steps were simple, such as taking shorter showers, letting the car stay dirty a while, and not watering lawns once the weather warmed. Even small things could help: fixing leaky faucets, turning off a running tap while washing dishes, not flushing ashes or cigarette butts. As winter approached, city hall threatened to cut pressure in water mains by New Year's Day if the drain continued. "This would divide New Yorkers into two classes—the people downstairs with water, and the people upstairs without," the Associated Press reported December 3. "Only two things—rain and rationing—can stave off the cut, city Water Commissioner Stephen J. Carney warned today."

Carney headed the city's grandly named Department of Water Supply, Gas and Electricity. His department began cooperating with others, assigning funds and nearly five hundred city inspectors to help cut water wastage. Health inspectors had the power to issue summons to people wasting water "on the ground that it was a menace to health." The fire commissioner even ordered a check of fire equipment to eliminate leaks "and is considering the use of river water for fighting fires near the waterfront." Three days after Carney's comment about water rationing, his chief engineer issued a warning that was even more dire. Unless plentiful rainfall replenished upstate reservoirs, Edward J. Clark said in an interview on WJZ radio, New York could become a "ghost city, with no power, no health facilities and no fire protection." He added that while "we don't expect nature to treat us that badly" and only a "cataclysmic change of nature could bring this about," the water situation nonetheless was becoming critical.

———————

The root of the problem was simple: in more than three hundred years, Manhattan had never drawn drinking water from the great river on its western banks. Residents for generations had called it the North River, but most knew it now as the Hudson. It is the largest and most majestic of several interconnected waterways that surround the city and empty into the Atlantic Ocean. Together, they constitute an estuary—a semi-enclosed ecosystem where sea tides meet freshwater.

Estuary water is too salty to drink. Even well upriver, north of the Tappan Zee Bridge at Tarrytown, the Hudson is "normally an estuary and typically has more ocean water than river water in the mix," a modern ocean-engineering expert writes. "Brackish water can actually work its way about 70 miles north of NYC to Poughkeepsie in droughts."

Early European settlers on Manhattan Island took their drinking water from streams, springs, and ponds. But human waste draining into gutters and industrial pollution later made that water less than potable. Consequently, New Yorkers have long depended on a complex municipal water system, the history of which is sprinkled with familiar names stretching all the way back to the first Dutch settlers. Director-General Peter Stuyvesant, for example, approved the digging of the city's first public well in 1658, along a street that is today called Broadway. Alexander Hamilton and Aaron Burr, unsurprisingly, clashed over a proposed dam and canal system in 1799. Nor should we omit William M. "Boss" Tweed, head of the infamous Tammany Hall political machine, who was appointed New York's first commissioner of public works in 1870.

Water consumption in New York City doubled three times between 1850 and 1950. By the mid-twentieth century its water system was vast and complex, ranging from sparkling upstate reservoirs to lowland water wells on Staten Island and Long Island. The scale was staggering. When the system was constructed in the early 1900s, it was considered "one of the most notable engineering enterprises ever undertaken." The Catskill Aqueduct alone—a ninety-two-mile-long underground conduit tying the mountains' watershed to city water pipes—was a construction feat "probably second only to the Panama Canal."

Needed expansion into the Delaware River watershed west of the Catskills, however, stalled during World War II for lack of building materials. Watershed and aqueduct systems expected to be available in the mid-1940s wouldn't be completed now until the mid-1950s. By 1946 water use had surpassed the existing system's safe minimum yield. Water Commissioner Carney assured New Yorkers that once construction of the $170 million Delaware Water System was finished "some time in 1952, the city will have a 'safe' supply until 1970—if use of water remains 'normal.'" Until then, however, the city's five boroughs largely relied on what experts called surface water supplies, from the Catskill and Croton watersheds.

Those two sources lay within New York State but seventy-five miles apart, on opposite sides of the Hudson River. The Croton watershed lay nearer to the city, in portions of Westchester, Putnam, and Dutchess Counties, east of the Hudson and west of the Connecticut state line. Croton River water first reached New York City by aqueduct in 1842, decades before the city tapped the Catskills' resources. By 1950 the Croton watershed covered 375 square miles and had a capacity of more than 100 billion gallons, stored in various lakes and basins, including the

Kensico Reservoir (elevation 357 feet) and the New Croton Reservoir (200 feet), both in Westchester County.

The Catskill watershed originated west of the Hudson a hundred miles above Manhattan and covered 571 square miles of mountains and valleys. It had two reservoirs: the Schoharie (elevation 1,130 feet) and the Ashokan (590 feet). They were connected via the eighteen-mile Shandaken Tunnel, which began at the Schoharie and fed into Esopus Creek a dozen miles above the Ashokan. The Shandaken Tunnel, too, was an engineering marvel, considered "one of the wonders of the world" thirty years earlier. Together, the two Catskill reservoirs provided a storage capacity of 150 billion gallons of freshwater. Its good mountain water filtered mainly through bluestone, with very little limestone to add bitter-tasting calcium. "As a result," says the *New York Times*, "New York has delicious bagels and pizza crust."

The Ashokan Reservoir in 1916, soon after its construction. *Courtesy of New York Public Library Digital Collections*

The Catskill watershed was connected to the Croton via the Catskill Aqueduct, which ran from the Ashokan to the Kensico Reservoir, then south to Yonkers, where it tied into the city mains. When conditions warranted, engineers could draw Catskill water into the Croton system or pump it from the Croton into the Catskill. At full capacity, the two watersheds together stored slightly more than 253 billion gallons. "It is therefore evident that the Catskill and Croton systems are really one system and should be considered as such in determining the dependable supply," a city engineer wrote. Newspapers often combined them and rounded up, referring inaccurately to New York's "1,000-square-mile watershed."

Exaggerations aside, the watershed system's capacity was impressive and invaluable. As a Hunter College geology professor wrote in 1959, "The large size of New York City's reservoirs, due both to nature's potentialities and to man's decisions, makes it possible for water to remain in them for a long time—in the Catskill reservoirs for as much as six months, in Kensico . . . for about three weeks." Construction on the vast, interconnected system would continue for many years, until by 1965 every important Catskill waterway but one had been dammed. "Sixty thousand acres of land had been appropriated, twenty towns and villages had disappeared, six thousand people were driven from their homes, and more than ten thousand graves had been dug up and the bodies moved to other sites," states a modern history of the area—all so that New Yorkers could enjoy potable water.

In 1950 city dwellers got about 70 percent of their freshwater from the Catskill portion of the system. But they knew the region best for its rustic history and vacation resorts. The area known as "the Catskills" was large and loosely defined. To mountain residents, it meant four New York counties: Ulster, Greene, Sullivan, and Delaware. The Catskill watershed, however, drained portions of Ulster, Greene, and Schoharie Counties. In the 1850s, Washington Irving wrote that the mountains "derived their name, in the time of the Dutch domination, from the catamounts by which they were infested." Roughly translated from the Dutch—Kaatskill or Katskill—the name meant Cat Creek. A century after Irving, truck and dairy farms still dotted the slopes and valleys.

The bigger draw for New Yorkers, however, was the area's vacation resorts and cottages. The first popular resort, the Catskill Mountain House, opened in 1824. The tourism boom began in earnest after the Civil War, in the 1870s and 1880s, and was still going in 1950, when Sullivan County alone had around six hundred hotels. Thousands of businesses of every size and type depended on summer visitors to make a profit for the year. Over recent decades, the resorts increasingly had attracted Jewish owners and guests. "Since the new Catskills were so closely tied to the Jewish resorts, they earned some nicknames, used both affectionately and derogatorily. 'Borscht Belt' was popular by the 1940s, and the 'Sour Cream Sierras' by the 1950s."

Tourism was vitally important to the economies of small Catskill communities such as Margaretville, seen here in 1947. *Delaware County Historical Society, photo by Bob Wyer*

As they neared the last days of 1949, New Yorkers worried not about tourism but about water. City hall slowly began devoting more attention to the drought, like a great ocean liner edging into the stream. The city's mayor at the time was William O'Dwyer, the one-hundredth person to hold the position. He was the quintessential New York politician: an immigrant born in County Mayo, Ireland, he'd been a city policeman on the waterfront beat, a lawyer turned magistrate, and a district attorney who had helped prosecute the gangsters of Murder Inc. Despite his gang-busting reputation, O'Dwyer had risen through the Tammany Hall organization that still dominated the city's Democratic politics, a fact that would complicate his legacy. He challenged and lost to incumbent mayor Fiorello H. La Guardia in 1941, then enlisted in the US Army during the war, investigating graft for the inspector general's office and rising from major to brigadier general. He ran successfully for mayor the second time in 1946.

"Throughout his later career, his warmest friends remained policemen and sports writers, and to both groups he was not known as Judge or General or Mayor, but simply as Bill-O," the *New York Times* recalled years later. "As mayor, O'Dwyer lacked the witty urbaneness of Jimmy Walker, or the excitable, explosive energy of Fiorello LaGuardia," the Associated Press likewise remembered. "But he managed to impart the feeling that he had the common touch, and the response was political promotion that carried him upward over the years."

Facing the growing water crisis, Mayor O'Dwyer asked city workers to report all evidence of waste. Water Department inspectors began posting fifty thousand posters that urged New Yorkers to stop wasting water. The Board of Transportation printed fourteen thousand scarlet-and-white posters for display in subway and elevated cars. The superintendent of schools mulled shutting down thirty-nine swimming pools. Using water to wash cars became an offense—a magistrate fined an unemployed Brooklyn man ten dollars for washing his—but Commissioner Carney held off ordering commercial car washes to close, saying he didn't want to cost people their jobs. Even among business owners, however, the use of constantly running water was forbidden, "either by

voluntary co-operation now or by enforced regulation in a few days."
(The owner of the Speedway Auto Wash in Rockaway Park, Queens,
promptly sank his own legal well, digging thirty-two feet into the sand
in four hours. "After that it was just a matter of connecting an electric
pump to the well and they were in business again.")

New York asked eighty upstate towns and villages that drew 45 million gallons of water a day from the city's system to cooperate with the
city conservation program. The drought didn't touch every community
situated above Manhattan, however. Four villages between the Hudson
and the Connecticut state line, southeast of Albany, declared they had
no water shortage at all. Dr. Samuel Johnson, chief executive of tiny
Kinderhook, went so far as to claim that his community had enough
water for "two or three villages." But Kinderhook's surplus wouldn't
have lasted a day in any neighborhood in Harlem or Queens. Mayor
O'Dwyer made the city's message clear. Speaking from a three-room
suite at Bellevue Hospital, where he was undergoing a thirteen-day rest
cure for exhaustion, a severe cold, and heart strain, he said, "We must
conserve water. We cannot take it for granted that nature will replenish
our rapidly dwindling supply. Therefore, it is necessary to prepare at
this time for whatever emergencies may arise."

The water situation grew so worrisome that Francis Cardinal Spellman asked worshippers in the city's 381 Roman Catholic parishes to
pray for rain. "The prelate ordered that the prayers be included in
all regular services for the next three months. Church officials said
the request was 'very unusual.'" One prayer was "Give us, we pray
Thee, O Lord, wholesome rain and graciously bathe the parched face
of the earth with heavenly floods." The chancellor of the archdiocese
also sent Catholics a letter reminding them of the critical shortage and
the "absolute need to avoid any waste of water." Protestant and Jewish
congregations across New York joined in both prayer and conservation.
"If you want to be able to take baths and live in a civilized way," an
Episcopal clergyman warned, "save water now."

The O'Dwyer administration proceeded with its more direct, secular approach, declaring New York City's first waterless holiday, during

which residents were encouraged to cut back on water use as much as possible. It took place on Friday, December 16; some newspapers dubbed it Dryday, while the United Press sent out a wire story under the byline Sahara-on-the-Subway. Ninety-five of 114 householders surveyed by telephone said they were complying with the mayor's behest. "My husband walked out this morning without a shave, and he looked like hell," one woman declared. Another told the caller, "I'm not even washing the baby. We're stacking the dishes to save for one washing tonight. We're not washing the clothes." A man on the Bowery said simply, "The boys don't plan to take no baths today."

The *New York Times* ran a photo of Commissioner Carney gazing into an office mirror while stroking his stubbly chin. Affixed to the mirror was a sticker printed with the city's new mantra: "STOP WASTING WATER. Your health depends upon it. Turn off fawcets securely. Don't let water run needlessly." Aides estimated that water use in the city on the first official dry day was 90 million gallons less than the day before— the lowest daily total in twenty-one years. Despite the conservation, water levels in the watershed reservoirs inched even lower. Weekly dry days would continue for months.

Much of the responsibility for managing New York's response to the crisis over the coming year would fall to Commissioner Carney. He would appear on every television station in New York, urging citizens to save water. "It is entirely possible that this same Steve Carney holds the record for having been seen and heard on TV more often than any other person over a short span of time," a TV writer later recalled. At the time, Carney was forty-one years old, educated in Brooklyn parochial schools, the son of a city firefighter who had died of injuries suffered in the line of duty. A prizewinning speed typist in his youth (second in an international championship), Carney was appointed educational director for the Woodstock Typewriter Co. in Manhattan at age nineteen. He entered Democratic electoral politics in 1935, becoming the youngest member of the Board of Aldermen at twenty-seven. Mayor O'Dwyer appointed him to his current powerful, $15,000-a-year post in July 1948. Fortunately for everyone, Carney was an intelligent and capable water commissioner. By late 1949 he may have begun to realize that the growing water crisis would be the defining challenge of his career.

———————

The Catskill and Croton reservoirs each received rain on Friday, December 23, and by Christmas Eve the situation looked a bit brighter. Overall water levels rose for the second time in a week, the first increases since June. "When New Yorkers hang up their stockings tonight, they may well ask Santa Claus to bring this drought-stricken city some more water," the Associated Press reported. "And there's just a chance the jolly old fellow may do it." The net water loss for the month was only 4 billion gallons—much better than the monthly average of 30 billion gallons since summer. "But now we hope we have touched bottom," said Chief Engineer Clark. The city's reservoirs on New Year's Eve stood at 35.9 percent of capacity—an awful figure but ticking upward from the previous water reading of 33.4 percent on December 12.

Nonetheless, authorities cautioned that there was no end in sight for the crisis without additional precipitation. Whether most realized it or not, drought-stricken New Yorkers needed storms over the Catskills. "If winter brings heavy rains and deep snows on the watersheds, the crisis may be postponed," the Newspaper Enterprise Association wire service reported. "Otherwise, the water experts warned New Yorkers, there is no way to relieve the emergency before 1952; the engineers have no magic that will turn on a new water supply before then."

More specifically, the city needed the storms that did appear over the Catskills to surrender a greater proportion of their moisture than they had done recently. If more precipitation fell, billions more gallons of freshwater would trickle down the green slopes, into the creeks that began small and grew steadily larger as they descended, until finally they entered the reservoirs and flowed through the Catskill Aqueduct and down into the city, to ease the shortage caused by the drought.

Put that way, the solution seemed almost simple. All the city needed to do was hire a rainmaker—a *scientific* rainmaker, certainly, but a rainmaker nonetheless. In the new postwar atomic age, anything seemed possible. After all, wasn't science already making snow? And what was snow but frozen rain?

2

Snow

The godfather of American rainmaking in the years immediately following World War II wasn't a meteorologist. Rather, it was Dr. Irving Langmuir, a Nobel Prize–winning chemist born in Brooklyn on January 31, 1881. By the time the New York drought made rainmaking a tempting option, Langmuir was working in Schenectady, about an hour north of the city by plane.

Langmuir had been "the kind of child the visiting aunts referred to as 'that inquisitive boy.'" As his family moved for his father's career in insurance, he attended school in Paris and then Philadelphia. He later earned a degree in metallurgical engineering from the School of Mines at Columbia University and a doctorate in physical chemistry from the University of Göttingen in Germany. (His older brother Arthur was also a respected chemist.) By the 1940s Langmuir was associate director of the General Electric Research Laboratory in Schenectady, which he had joined in 1909.

Although he didn't teach at an Ivy League university, in many ways Langmuir fit the stereotype of the brilliant but absentminded professor. Writer and historian Ginger Strand recounts how Langmuir could pack a blue serge suit instead of outdoor gear for a sailing trip, leave a yellow trail of footprints after stepping in a can of paint, and pass by colleagues without even a nod of recognition, including one who had fallen on the stairs. Yet he was also a civic-minded husband and father. Langmuir had organized one of the first Boy Scout troops in Schenectady and led it as scoutmaster. He was a great outdoorsman who had climbed peaks in the Alps as a teenager and "was exploring mountains on skis long before most Americans knew what skis were." He had owned and flown an open-cockpit Waco airplane in the 1930s but once astonished Charles Lindbergh by saying he would rather ski than fly.

Born in the nineteenth century, Langmuir was a man of the twentieth, a modern, intuitive scientist with an "uncanny ability to turn observations to useful account." By 1920 he had developed the gas-filled lightbulb, which "raised lighting levels in workshops, schools, offices, and homes, and made possible the projector for home movies," plus a condensation pump that produced the highest vacuum ever created. "His outstanding accomplishments in gas reactions, the working of tungsten and the development of the submarine detector during the [first world] war, have won him international fame," the *Brooklyn Eagle* proudly reported. He went on to develop a compact, 20-kilowatt "super tube" for wireless radio, which Nobel-winning radio pioneer Guglielmo Marconi considered "the greatest development of the age." What Marconi mistook in 1922 for wireless signals emanating from Mars was only Langmuir, "sitting at the transmitting key of his 150,000 meter wave length experimental set . . . 'making dashes' of regular intervals in the little upstate city [Schenectady]. We have yet to hear from Mars."

Langmuir later developed atomic hydrogen welding, which made it possible to weld certain metals for the first time, as well as to weld extremely thin sheets. He also experimented with oil films on water, discovering "an entirely new branch of chemistry, known as two-dimensional or surface chemistry, in which phenomena are found entirely different from any known before." Langmuir received his own Nobel Prize for chemistry in 1932; he was only the second American and the first industrial chemist to receive one in the discipline. By 1937 he had published 167 papers and held a slew of patents. His former boss at the research lab described him as a man "who continually embarks upon mental voyages in regions so nearly airless that only the mind can

Nobel laureates Irving Langmuir (left) and Guglielmo Marconi compare notes in a lab. *Courtesy of Library of Congress*

breathe in comfort." Langmuir himself summed up his philosophy in a 1941 article for *Boys' Life*, the scouting magazine, published on the eve of America's entry into World War II. "Train yourselves. Don't wait to be fed knowledge out of a book. Get out and seek it," he wrote. "Make explorations. Do your own research work. Train your hands and your mind. Become curious. Invent your own problems and solve them. . . . Seek out answers to your own questions."

In his sixties by this point, Langmuir continued his frenetic pace before and during the war. With his brilliant GE protégé, Vincent J. Schaefer, he conducted research for the Chemical Warfare Service on filters used in gas masks. Years later, Schaefer recalled that his mentor "approached a problem with the zeal of a Crusader, never knowing when to quit so long as an unexplored avenue beckoned." Their filtration work led them to develop a smoke generator used to hide troop movements. They tested the prototype, nicknamed Junior, in June 1942 near a promontory in the Schoharie Valley called Vroman's Nose. The device proved several hundred times better than smoke pots for creating a screen—and effectively blocked the view of everything for three or four miles up the valley.

By spring 1945 Allied forces in Europe were using GE's new M1 smoke generators to form a "mystery wall," shielding the movements of the Canadian First, British Second, and American Ninth Armies. Troops generated the screen daily, from an hour before first light until an hour after dark. "The smoke screen which has baffled the Germans starts along the west bank of the Waal river near Nijmegen," the United Press reported. "It extends from there to where the Waal joins the Rhine near Millingen and thence along the west bank of the Rhine, a total distance of 66 miles."

Back in Schenectady, the clouds of war called for more than just billows of smoke. Weather was an important factor in warfare, so the GE Research Lab studied cloud formations. Langmuir's team delved into aircraft icing (a particular problem in the Aleutian Islands of Alaska), icing nuclei, and cloud physics. Special research for the US Army Air Forces often took Schaefer to the observatory on New Hampshire's Mount Washington, the site of some of the highest winds and fiercest weather in the country. There he developed several meteorological instruments, including a cloud meter to measure water content in clouds. Schaefer also studied the troublesome radio static produced when aircraft flew through snowstorms. All of this was important to the Allied war effort.

Once the fighting ended, Langmuir and Schaefer continued their weather research. They couldn't have known in 1945 just how important and promising their work would become—although Langmuir, his eyes turned more toward the future than a floor or stairway right in front of him, might have had an inkling. Wherever the Nobel laureate led, exciting developments tended to follow. A meteorological professor later dubbed him "Langmuir the Indomitable."

The background of Langmuir's protégé was modest by comparison. Vince Schaefer was born on July 4, 1906. He had a scientific mind, and one of his earliest memories was of seeing Halley's Comet in 1910. As an adult, however, he had neither a "Dr." before his name nor any prestigious initials after it. He had dropped out of Schenectady High School at sixteen to seek work; his parents were ailing and needed his help to support the family. Vince joined an apprentice course for machinists at the big company locals called "the GE," finishing his apprenticeship in the GE Research Lab machine shop. He left for a time to work trimming trees but soon returned as a model maker for the scientists and began working his way up in the lab.

Extremely skilled with his hands, Schaefer was equally adept at using his mind. His home bookshelves creaked under the weight of over a thousand volumes, which provided him an excellent technical background. Schaefer held on to his job through the early years of the Depression, when many workers were losing theirs. While a self-made man, he wasn't entirely self-educated. Schaefer referred to his time working beside the Nobel laureate as "Twenty Years at Langmuir University." (Decades later, the phrase would become the subtitle of a posthumously published autobiography.) Langmuir made Schaefer his lab assistant in 1932, shortly after returning from Sweden with the Nobel Prize. Schaefer's mother later wrote that Vince could have gone to college in his late twenties but didn't want to lose the research time. By 1938 he was a member of the American Chemical Society. Schaefer called Langmuir "the Boss" and would work with him until the chemist retired.

As oblivious as Langmuir was in some aspects of his life, he clearly recognized a kindred spirit in Schaefer. The scientist and machinist shared, for instance, an enduring love of the outdoors. Schaefer had organized the Mohawk Valley Hiking Club and the Schenectady Wintersports Club while still in his twenties. "In his work and in his diversions he observes the things others don't see," Langmuir said of him while presenting a civic service award at the Schenectady Elks club in 1941. "That is one of the things that makes him a good research worker. He has a wonderful curiosity, which he puts to the best possible advantage."

By 1946 Schaefer was Langmuir's right-hand man. The University of Notre Dame awarded him an honorary doctor of science degree in June 1948, the first of three honorary doctorates the high school dropout would receive. "Too frequently the awarding of honorary degrees by colleges is overworked, but the honor paid to Mr. Schaefer is one with which no one is likely to take issue," applauded the hometown *Schenectady Gazette*. The researcher led an equally fulfilling life outside the lab. Schaefer had a wife and three children and a winning personality: warm, good-humored, supportive, endlessly curious, and inventive. "It didn't take long to like Vince," a lifelong friend recalled.

The story of Schaefer's remarkable rise and his relationship with Langmuir was so compelling that General Electric eventually created a print ad around it, for college publications such as the Union College *Concordiensis* in Schenectady. "Stories like this are possible where emphasis on research and incentives for creative thinking are the tradition," the text proclaimed. "By 'finding' men of high caliber, General Electric stays in the forefront of scientific and engineering development."

Because of the many astonishing things that happened there, people often called the GE Research Lab the "House of Magic." The researchers themselves disliked the nickname, being serious scientists and not mere magicians. Schaefer later wrote that in some ways the lab was "perhaps superior to the best universities, since technicians were readily available to cope with problems of fabricating instruments and apparatus, while the new discovery which had practical application could often be followed to its ultimate use." Schaefer had begun as one of those technicians. "Well-trained men wouldn't do the things I do," he once said. "If I found anything, it's most likely because I didn't know better."

Schaefer had an almost childlike love of snowflakes, although his interest in the science of snow and rain evolved slowly. In 1940 he developed a way to make perfect impressions of snowflakes on a quick-drying plastic film. He quickly became known as the "snowflake scientist." While working on Mount Washington during World War II, he became fascinated by the natural phenomenon of supercooled clouds. In such a cloud, water droplets remain liquid, even when temperatures within the cloud itself are well below freezing. But the moisture freezes instantly when it contacts metal; this was the phenomenon behind dangerous icing on military and civilian aircraft.

Schaefer continued his study of supercooled clouds after the war. During summer 1946 he conducted cloud tests at the House of Magic, in a 6-cubic-meter home freezer made by GE. The units were hard to come by for regular homeowners in Schenectady and other towns in the years just after the war, but Schaefer wasn't using his to store steaks and ice cream. He lined the sides with black velvet and rigged lights so he could more easily see what was happening during his experiments. Schaefer badly wanted to find out why supercooled water droplets didn't turn into ice or snow.

Researchers knew that snowflakes formed around tiny particles that served as nuclei, like a grain of sand at the center of a pearl. Japanese physicist Ukichiro Nakaya had even synthesized snowflakes on the tips of rabbit fur in 1936. Schaefer tried more than a hundred times to form ice crystals around all sorts of minute particles, including carbon, sand, graphite, oil, sulfur, magnesium oxide, silica, volcanic dust, talcum powder, and even kitchen cleansers. Nothing worked. Then one sizzling summer day, July 12, 1946, he lowered a piece of dry ice (solid carbon dioxide) into his freezer, just to keep it cold. Within ten seconds, the freezer filled with ice crystals known as "diamond dust." This was the "fortunate accident" that Langmuir had believed in and exploited for years, the sort of serendipity "which in his definition consisted of the art of profiting from unexpected occurrences." Schaefer told his GE colleagues, "Now I know how to make snow."

Further research revealed that it wasn't the dry ice itself that created the ice crystals but the very low temperatures surrounding it. Droplets froze spontaneously at anything below –40 degrees Celsius, without any nuclei. (The temperature in Fahrenheit is also –40 degrees.) A needle or copper rod dipped in liquid air had the same effect, as Schaefer liked to demonstrate with a flourish. Langmuir was thrilled that the whole thing was so simple that high school students could replicate it. "It is a wonderful experiment," the Nobel laureate said. "The effects to be seen are wonderful to look at and it is a simple matter to duplicate all the natural conditions of an actual cloud in the sky."

General Electric made a short film, "featuring Vincent Schaefer 'The Snow Man,'" to demonstrate his snowmaking technique. You can watch it online today. Schaefer scratches a bit of dry ice so that tiny fragments fall into the freezer. Streaks resembling tiny vapor trails immediately form, consisting of many millions of snow crystals that are far too small to see. "The particles grow very fast," Schaefer casually explains in a voiceover, as if talking to a group of schoolkids on a field trip. "They grow about a billionfold in volume in a few seconds. And if it were possible to keep them supported in the air they would grow very large, just like the snow crystals that fall from the sky on a cold winter's day."

Schaefer and his freezer quickly entered scientific and meteorological lore. Photos of him leaning into the big, flip-top appliance—perhaps the same model that people had in their own basements or pantries— appeared in publications across the country, and would continue to do so for years. The scientist submitted his findings to *Science* magazine in the fall of 1946. He wrote that he would next try converting a super-cooled cloud to an ice-crystal cloud—not in a lab but in nature, "by scattering small fragments of dry ice into the cloud from a plane. It is believed that such an operation is practical and economically feasible and that extensive cloud systems can be modified in this way." Schaefer had already done it by the time the article reached print.

3

Mount Greylock

Vince Schaefer and pilot Curt Talbot, head of GE's flight test center, made the first practice snowmaking flight on Tuesday, November 12, 1946. It produced no results, because conditions weren't right. They tried again the next day—the same day that General Electric released news of Schaefer's groundbreaking freezer experiments four months earlier.

The pair lifted off from Schenectady County Airport at 9:30 AM in a rented single-engine Fairchild cabin airplane. Talbot dipped a wing and turned east toward the Berkshires in neighboring Massachusetts. Their destination was Mount Greylock, the highest peak in the state, rising nearly 3,500 feet in the northwest corner of the commonwealth. Talbot coaxed the little plane up to 14,000 feet as they approached a stratus cloud formation over the mountain. The temperature was 43 degrees Fahrenheit back in Schenectady but –4 degrees up here. This meant the cloud was supercooled. Approaching the base of the cloud, Schaefer also saw a brilliant corona around the sun and gorgeous iridescent colors in the formation itself, all produced by water droplets. These were further indications that test conditions were ideal.

Schaefer released tiny pellets of dry ice into the cloud with a small motorized dispenser set into the bottom of the plane. After 3 pounds had fallen through, "I looked to the rear and was thrilled to see long draperies of snow falling from the cloud we had just passed through," he later wrote. "I shouted to Curt and he swung the plane around so we could both see it." The electric dry ice dispenser then jammed. Exhausted from working at high altitude without oxygen, Schaefer opened the window and let the suction from the airstream pull the remaining 3 pounds of dry ice from its cardboard box. Thus, they dropped 6 pounds overall, along a three-mile track. Schaefer saw sunshine glinting off snow crystals. He turned to shake Talbot's hand and said, "We did it."

Within two minutes of Schaefer's radio signal that he was ready to start dropping dry ice, "a radical modification of the cloud took place and streamers of snow began to pour out of the base of the cloud," Irving Langmuir said afterward. "Though it was more than 50 miles away, we could easily see all this with the naked eye." Schaefer himself wrote in his lab notebook that "the rapidity with which the CO_2 dispensed from the window seemed to affect the cloud was amazing. It seemed as though it almost exploded, the effect was so widespread and rapid." "This is history," Langmuir told him when the plane landed.

The flight is considered a landmark today, and Schaefer's picture and an account of it appeared in *Life* magazine in the last issue of 1946. "That simple cloud seeding experiment was the first successful demonstration that a natural supercooled cloud could be converted at will into a cloud of ice crystals," one of Schaefer's many friends and colleagues wrote in a weather publication decades later. "The modern science of cloud physics and experimental meteorology had begun."

The only complication was that none of the snow streaming from the base of the cloud over Mount Greylock ever touched the ground. It fell two or three thousand feet before evaporating in drier air below. Lawrence Drake, a meteorologist skeptical of the GE experiments, later dismissed it as "a snowstorm that never took place"—a largely media-generated phenomenon. Langmuir, however, was thrilled. "In a five-hour flight, a single plane could generate hundreds of millions of tons of snow," he told journalists. "However, this would not necessarily mean a fall of more than a few inches on the ground." Still, this extra snowfall would melt in the spring, adding to the natural flow into irrigation and water projects.

"Man-made snow, every bit as real as that which makes for a 'white Christmas,' has been produced for the first time by Vincent J. Schaefer, a scientist of the General Electric Research Laboratory," a GE press release announced. Newspapers all across the country carried follow-up articles, often on the front page. Magazine coverage followed, *Life* calling Schaefer's effort a "near magical experiment." The science editor of *American Weekly* hailed the "sensational conquest over nature." An upstate editorial writer in Rome, New York, however, recalled other ballyhooed scientific experiments over the years that had led nowhere, and asked, "In any event, why should anybody want to cause a snow storm except the keepers of a few winter resorts? . . . We Romans would certainly not wish for any greater snowfall than we normally have."

Many readers simply didn't know what to make of the whole thing. Howard Blakeslee, a science writer for the Associated Press, wrote that Schaefer's first snow over Mount Greylock "carried a hint that a lot of study will be needed before man can produce snowstorms where and when he will." Blakeslee noted that although the snow Schaefer had produced consisted of tiny ice crystals, it was likely that larger snowflakes were possible. "The snow fell into dry air and all of it evaporated before it reached the ground," he wrote. "Under some other conditions such snow pellets might melt into raindrops."

Schaefer continued going up in the little Fairchild to see if he could produce a proper snowstorm. Flights on November 23 and 29 produced brief snow from the base of isolated, cumulus-type clouds. The team decided that such clouds weren't the right targets, because they formed and disappeared too quickly. The situation was different on the morning of December 20. The entire sky was overcast, with the US Weather Bureau predicting snow that evening. Schaefer went up around noon and began dropping 25 pounds of granulated dry ice at 7,000 to 8,500 feet, a thousand or so feet above the overcast. He also discharged 2 pounds of liquid CO_2 into the ragged gray layers below. The plane made four seeding runs altogether, then dipped down to 4,000 feet so Schaefer could check what was happening. He saw snow.

The snowfall began about a quarter past two in Schenectady and elsewhere throughout the region. Snow kept falling, piling up an inch an hour, through nightfall and beyond ten o'clock—the biggest storm all winter. "While the seeding group did not assume it had caused this snowstorm," General Electric reported, "it did believe that, with weather conditions as they were, they could have started a general snowstorm two to four hours before it actually occurred, if they had been able to seed above the clouds during the early morning." Nobody really knew whether GE had made it snow, but Langmuir and his team of weather wizards soon concluded that they had. Schaefer doubted that he had done more than cause a few extra inches along a narrow band, but "the subsequent storm was a big one that resulted in strict orders from the GE Legal Department to cease all such activities!"

Schaefer "continued to skywrite weather history in bold strokes," the *Saturday Evening Post* reported the next autumn. "Succeeding flights produced snow that hit the ground. And citizens of Schenectady began to look at every snowflake as evidence that 'that man Schaefer must be up there stirring the clouds again.'" Schaefer mildly objected to the publicity that swirled around him like winter flurries. "I have produced some snow but don't want to be known as a snowmaker," he said later. "I deal in experimental meteorology."

———

The third member of the GE Research Lab's snowmaking triumvirate was Dr. Bernard Vonnegut. Like Langmuir and Schaefer, he wasn't a meteorologist. He had earned his doctorate in physical chemistry in 1939 from the Massachusetts Institute of Technology. Vonnegut first worked in industry, researching the surface properties and breaking points of glass. He became a civilian researcher studying the important aircraft-icing problem during the war, sometimes flying on a B-24 Liberator bomber used as a test bed. He also looked into the adhesion of ice and the properties of supercooled clouds.

Vonnegut joined General Electric as the war was winding down. His younger brother was an army infantryman who had been captured during the Battle of the Bulge and then survived the firebombing of Dresden. Kurt Vonnegut later would work for GE as well, taking a job in 1947 as a writer in the GE News Bureau to support his growing family. Bernard's work would inspire Kurt to create the substance ice-nine in his 1963 science fiction novel *Cat's Cradle*. He based one of the novel's characters, scientist Felix Hoenikker, on Irving Langmuir.

The elder Vonnegut brother began his job at the research lab in the chemistry group, researching the supercooling of liquid tin. He shifted to cloud seeding and cloud physics in mid-1946. Schaefer had met Vonnegut through his work with the deicing laboratory at MIT during the war and was delighted to welcome him to Schenectady.

The GE weather wizards wondered what else besides Schaefer's dry ice might produce ice crystals in a supercooled cloud. So Vonnegut began poring through pages of tables in a handbook of X-ray crystallography, looking for suitable materials. He narrowed nearly two thousand choices down to a manageable few. He and Schaefer began testing them in the famous freezer. Vonnegut eventually focused on silver iodide (AgI), an odorless yellow powder commonly used as an antiseptic and in photographic emulsions. Despite the name, it was inexpensive. When Vonnegut introduced a vaporized form into the freezer, silver iodide formed ice crystals as readily as dry ice had, and the effect lasted much longer. This was November 14, 1946—the day after Schaefer's first successful snow flight over Mount Greylock. The team knew now that Vonnegut was on to something.

Schaefer and Vonnegut had come across different processes that had nearly the same effect. "Dry ice could make crystals of ice form by lowering the temperature to that necessary for nucleation on whatever might be available as nuclei, or even in the absence of any nuclei," a history of weather modification later stated, "while silver iodide provided a nucleus that was much more efficient than those occurring naturally." Vonnegut's silver iodide particles, however, persisted much longer in the air than Schaefer's finely granulated dry ice. The scientist wandered through his neighborhood on crisp, clear winter nights, making ice-crystal fog by burning newspapers impregnated with silver iodide in a small oil burner. The tiny particles floated so long, the *Saturday Evening Post* related, that when Vonnegut fired up a newly devised silver iodide generator on the roof of the GE Research Lab in Schenectady, "they rode the breezes and started a small snowstorm in a grocer's frozen-food cabinet in suburban Alplaus, six miles away."

Langmuir, Schaefer, and Vonnegut jointly presented GE's findings at a meeting of the American Physical Society at Columbia University in late January 1947. "While the work has not yet progressed far enough to permit predicting actual uses, the discovery . . . might make possible the 'seeding' of vast areas of atmosphere with foreign nuclei," the *New York Times* reported. Since artificial nuclei don't melt or evaporate like ice nuclei, they "might be introduced by generators on the ground into the atmosphere and remain there until they produce snow." The best seeding material, the scientists agreed, was silver iodide. The *Schenectady Gazette* later added, "As explained by Dr. Vonnegut, crystalline particles of silver iodide so closely resemble the structure of an ice crystal that supercooled water droplets in a cloud are literally 'fooled' into using the particles as nuclei and thus turning to snow."

Vonnegut later penned an article for the International News Service (likely helped by the GE News Bureau) in which he described a simple smoke generator that could vaporize silver iodide, which would condense into particles less than 0.000001 inches in diameter. "In one minute, the generator produces more particles than grains of wheat are grown in the world in a year," he explained. The dry ice and silver iodide methods were difficult to compare, Vonnegut added, but laboratory tests suggested that they might supplement one another. He also quoted Langmuir's speculation that with large-scale silver iodide seeding from the ground, "it might be possible to alter the nature of the general cloud formations over the northern part of the United States during the wintertime."

Once he had perfected the device, General Electric explained Vonnegut's generator in everyday terms for consumers. "Fiercely-burning charcoal impregnated with a silver iodide solution emits thousands of sparks, each of which produces millions of silver iodide particles," a press release stated. "In the sky, the particles serve as nuclei upon which super-cooled or below-freezing water droplets in a cloud crystallize into snow. The snow then may turn to rain, dependent upon temperature and humidity of atmosphere near the ground." The GE News Bureau's copywriters dubbed Vonnegut's silver iodide method "rain by fire."

4

Possibilities

Irving Langmuir retired from General Electric on January 2, 1950—
the same day that William O'Dwyer began his second term as mayor
of New York in the midst of the drought emergency. Mayor O'Dwyer
observed the occasion with a simple ceremony at city hall and "a plea
for divine guidance to aid him in serving the best interests of the city."
He nearly hadn't sought reelection. "His renomination came only after
he had publicly declined to run, refused to hearken to a draft movement,
and then made a lightning switch just as the party leaders seemed scheduled to pick someone else," the *New York Times* recalled months later.

Friends worried that O'Dwyer's continued poor health and mounting political pressures still might force his resignation. "My doctor told
me I would not live another spring if I tried to stick it out for my second term," the mayor said later. After leaving the hospital the previous
month, widower O'Dwyer had married former fashion model Elizabeth
Sloan Simpson on December 20. The couple planned to leave New
York in mid-January for a Florida vacation—an unusual way to begin
a term as mayor of the country's largest city while it was in the middle
of a water crisis.

Dr. Langmuir, meanwhile, was eager to get back to work despite
his retirement. The scientist considered it only a technicality, since he
remained a consultant for GE, primarily on a federal weather contract

known as Project Cirrus. (So named, by Vince Schaefer, because cloud seeding creates the same sort of ice crystals that constitute high, wispy cirrus clouds, sometimes called horsetails.) "It would be hard to imagine this Schenectadian, regarded as one of the greatest scientists of modern times, doing else but continuing to work regardless of how he was listed at General Electric," the local newspaper observed. Legendary gossip columnist Walter Winchell scooped everyone by announcing what the scientist would say in his next big speech: "At the Astor on the 25th Dr. I. Langmuir (Gen. Electric rainmaker) will address the American Meteorological Society. . . . The Nobel Prize winner will startle weathermen with: 'Anyone with very little equipment on the ground (and at moderate cost) can produce rain IF—super-cooled cumulus clouds are present.'"

The week before Langmuir's planned rainmaking speech, New York took an extraordinary step to deal with the water emergency should the drought worsen. City hall prepared to tap the Hudson River miles upstream, regardless of whether other cities in the area approved. "New York public sentiment probably would be against drinking water from the Hudson River," Charles G. Bennett of the *New York Times* had written before Christmas. "But then, if New Yorkers will waste good Catskill-Croton water at the rate of 200,000,000 gallons a day, they may not be in a position to protest too much wherever their water comes from." The city administration sought and received state approval, so it wasn't outright theft, and the city hadn't yet built the necessary pumping station. The message was clear nonetheless: New York's five boroughs mattered more than smaller communities lining the banks of the great river.

"Water starved New York City today was granted permission to pump 100,000,000 gallons daily from the Hudson River near Chelsea as an emergency supplement for the depleted metropolitan supply," the *Brooklyn Eagle* reported on January 14. Situated on the east bank of the river ten miles below Poughkeepsie, the Dutchess County hamlet of Chelsea was to be the site of a planned $5 million pumping station. It would be capable of sending water to the Kensico Reservoir via the

Delaware Aqueduct, but New York City officials promised to use it only out of "dire necessity." Nonetheless, Dutchess County vehemently objected; it feared that siphoning water from the Hudson might affect the flow of its streams.

The city of Poughkeepsie also protested the plan. It drew all its water from the Hudson, pumping continuously. In October it had come within minutes of running entirely out of water. Poughkeepsie mayor Horace Graham said New York's pumping plan threatened "the safety, health and welfare" of half a million residents. His smaller city's water supply, he added, "would be endangered if the salt water mass moved up the river as a result of New York City's tapping the Hudson for such a large quantity daily."

Republican governor Thomas Dewey supported the Chelsea pumping station, and the Republican-controlled State Water Power and Control Commission granted New York City permission to build it. But the commissioners weren't happy about it. "Rapping the city for 'failure since 1931 to prepare for the normal increase in needs,' the agency warned local officials that it is their 'solemn duty' to develop other emergency sources of supply immediately," the *Long Island Star-Journal* reported. Commissioners also warned that New York's current emergency measures "will not solve the problem," adding, "It is the city's solemn duty to develop other emergency sources immediately."

Permission to build and use the Chelsea pumping station extended only until January 1, 1957. The *New York Times* reported "the Hudson River work would be done by late next summer. Officials have indicated that the Hudson water would be used only in extreme emergencies." Against this dramatic backdrop, Dr. Langmuir prepared to address the AMS meeting at the Hotel Astor.

Exactly one year earlier, in 1949, Langmuir and the US Weather Bureau had politely clashed over the economic importance of rainmaking. In a morning session at the same hotel, Langmuir and Vincent Schaefer had declared that Project Cirrus produced "positive indications of cloud modification or conversion of supercooled clouds to ice or snow crystals." In some mountainous areas, Langmuir said, annual rainfall might be doubled, especially in tropical countries such as Honduras and Costa Rica. He acknowledged, however, "You can't make rain from any old cloud, any old time, any old place."

Later on that same day, five Weather Bureau researchers had told attendees that recent government experiments in Ohio hadn't produced any indication that seeding clouds with dry ice actually could induce what they called "self-propagating storms." Efforts to make rain or control the weather, they held, were "of relatively little economic importance." The bureau researchers frankly doubted Langmuir's earlier assertion that a single 0.375-inch pellet of dry ice had opened a hole in a cloud a mile wide. During a luncheon speech, Dr. Francis W. Reichelderfer, head of the bureau, said that weather events were fantastically complicated, to the point that humans were unlikely to have much effect on their natural patterns.

Reichelderfer had emerged as Langmuir's greatest doubter and critic. Then fifty-three years old, he was greatly respected within the meteorological community, where people called him "the Chief." He had studied in London, Berlin, and Paris, and earned his master's degree from the University of Bergen under pioneering meteorologist Jacob "Jack" Bjerknes, studying how frontal systems and air masses reacted to each other. But the Chief was no hidebound traditionalist.

He was a midwesterner who had attended Northwestern University, studied advanced meteorology at Harvard University's Blue Hill Observatory, and earned a commission in the US Navy during World War I. "When Dr. Reichelderfer entered Government services . . . as a Navy meteorologist, weathermen were still flying kites to measure atmospheric temperatures and humidity." He earned his wings as a naval aviator and served on the navy's big, silvery airships, with a stint as an observer on the ill-fated German airship *Hindenburg*. He also charted weather conditions for flier Charles Lindbergh and won his praise. Reichelderfer was a navy commander when President

Franklin Roosevelt named him head of the Weather Bureau in 1938. He received his doctorate in science from Northwestern University the following year.

A decade later, he was a member of the National Academy of Sciences and the National Advisory Committee for Aeronautics, and a former president of the American Meteorological Society. He considered weather forecasting a "fascinating game" but said his job as head of the bureau was "keeping the machinery running." During the January 1949 AMS meeting in New York, Reichelderfer and his scientists recommended devoting more research to developing accurate forecasts and placing less emphasis on trying to change the "unbenevolent features of the weather." Reichelderfer believed in conducting fundamental meteorological research, whether for air navigation, farming, or other weather-dependent activities.

This view fundamentally differed from Irving Langmuir's. The GE scientist believed, to the contrary, that many natural phenomena were so complex, and triggered by such tiny variables, that they were utterly unpredictable. Today we call this the butterfly effect: that a butterfly flapping its wings in China can cause a hurricane on another continent. Langmuir thought something similar applied not only to science but also to history. "I believe that it is impossible to predict the future," he would say later in 1950, "because so many small things may influence the men who make history—things like their health; the way they might feel at a certain time." As for weather, the flamboyant Nobel laureate would later declare, "It is easier to make the weather than it is to forecast it."

Even in his criticisms of Langmuir, Reichelderfer took a much more measured and cautious approach. "The Weather Bureau has not said that rainmaking is impossible; we have said simply that the circumstances where artificial methods of inducing precipitation are of practical value apparently occur very infrequently," he later wrote to the *New York Times*. Asked during the 1949 AMS meeting about the difference of opinion between his bureau and General Electric, Reichelderfer replied simply, "No one knows the answer." Langmuir, as always, considered him and his bureau to be far too hesitant and rule-bound. Their public disagreement in 1949 had captured some newspaper space, but the meeting this year, in 1950, would produce much bigger sparks and headlines.

———————

Langmuir's 1950 presentation was a blockbuster, perfectly timed and delivered in the right city for maximum attention. Columnist Winchell hadn't come close to predicting its full effect. The United Press afterward called Langmuir's talk a "dramatic play-by-play description of a man-made rain storm that produced enough water to fill all of New York city's reservoirs and then some."

The GE scientist described an experiment the previous July near Albuquerque, New Mexico, in which scientists had seeded supercooled cumulus clouds with silver iodide smoke generated from the ground. The generator ran for eighteen hours and used 0.667 pounds of silver iodide, costing just twenty dollars. The result, Langmuir said, was a deluge: 320 billion gallons of rain. An earlier experiment, in October 1948, had brought down 160 billion gallons.

"Dr. Langmuir holds that silver iodide smoke or dry ice can make useful rain," wrote Howard Blakeslee, the AP science writer. "The U.S. Weather Bureau has reported the rains so far made were not useful, and might have fallen anyhow." But the bureau's qualms often appeared near the bottom of newspaper articles about Langmuir's speech. The Nobel laureate himself thought it "highly probable" that modern rainmaking techniques could help New York City authorities end the water shortage. "They ought to hire a crackerjack meteorologist," Langmuir said. "We will give him the fundamental information from our research. Let him go to work and I think he could lick the shortage."

General Electric itself was sticking to research and wasn't about to sign a rainmaking contract with New York City or anyplace else. "We can't go into the professional application of making rain, because we still have many problems to solve in the study of basic cloud physics," Vince Schaefer had said in December. "We don't know any way to make it rain without the right type of cloud." Indeed, he later declared, "I've never made a drop of rain in my life, and I don't intend to. What we're primarily interested in at the General Electric laboratory is what causes rain. When we learn that we will have the key to many of the mysteries of the atmosphere."

But, like Langmuir, Schaefer did believe that cloud seeding might help to ease New York's drought emergency. He had told a United Nations conference in August about the "future economic benefits" that dry ice and silver iodide seeding might offer. "The new science, he said, might also be used to increase the snowfall in mountainous country, thereby providing a more abundant water supply to streams fed by moisture from the mountains," the United Press reported.

By the beginning of 1950, New York City had quietly completed a three-month study of artificial rainmaking. The head of the city's Law Department, the corporation counsel, hadn't yet provided an opinion on the data, but Commissioner Carney had decided against experimenting with silver iodide, citing lack of proof that it worked. Now, however, Langmuir's report on the New Mexico tests—and perhaps the negative reaction to the Chelsea pumping plan—prompted Carney to take a second look. As the *New York Times* noted in an editorial, "Dr. Langmuir is no rain-making crank, but a Nobel Prize winner who ought to be consulted in the present extremity."

On February 10, the Associated Press reported, "New York City turned seriously—if dubiously— . . . to the possibility of artificial rain-making." Carney maintained that the issue was still "very controversial" but added, "We are not closing our minds to the possibility of getting rain from any source whatsoever." He arranged to meet with Langmuir in Schenectady the following week. The scientist thought he could reverse the bureaucrat's skepticism. "Mr. Carney evidently does not understand what has been written," Langmuir said, "and I will simply attempt to explain it to him."

The commissioner and chief water engineer Edward Clark visited Langmuir on Wednesday, February 15. The pair arrived that morning from Albany, where they had stayed overnight after discussing the city's water problem with state legislators. While still cautious about cloud seeding, Carney said in Schenectady, "I don't want to leave a stone unturned." Langmuir had spoken to a local Boy Scout troop and their fathers the previous evening, predicting that "to a large extent" science would largely control weather in America within ten years, and that the government would "allot rainfalls in sections of the country where conditions are favorable." Reporters didn't relate whether he repeated those sentiments to the water commissioner.

Carney and Clark spent the day with GE's rainmaking triumvirate— Langmuir, Schaefer, and Vonnegut—in the research lab and at Langmuir's home. Schaefer showed the visitors the famous freezer. This wasn't General Electric's first dog and pony show, and everything went perfectly. (A later visit by US senator Joseph C. O'Mahoney of Wyoming, in contrast, would go awry when he got stuck in an elevator with a GE vice president. "Is there an engineer in the house?" O'Mahoney joked.) Ginger Strand has captured the company's success in entertaining and informing the visiting New Yorkers in her book *The Brothers Vonnegut*: "As always, the News Bureau was there to get photographs, and though Carney had a long, mournful face and Clark wore a skeptical frown, at the end of the day, the two men met with a large group of assembled reporters at the Hotel Van Curler and made a startling announcement."

"I am certainly going to suggest to Mayor O'Dwyer the appointment of a skilled, technically trained man to study the feasibility of artificial rainmaking for New York City," Commissioner Carney said. "Just as we have a traffic expert, New York City needs a man to study the weather problem and advise officials what can be done to ease the situation."

Langmuir urged New York to appoint "a competent meteorologist who has nothing to sell and who will study and plan the project with a conservative approach . . . a meteorologist of the new school of thinking." Carney realized that the city would have to start with basic research and data-gathering, but his reservations about the science itself had vanished. After spending a day with Langmuir, he had become a believer. "I am very, very enthusiastic," he said, "about the possibilities of producing artificial rain."

5

Crackerjack

There was little or no discussion between the GE scientists and the city officials about whom Langmuir's competent non-salesman might be. The Nobel laureate passed along a name suggested by Vince Schaefer: Dr. Wallace Howell, a thirty-five-year-old research meteorologist at Harvard's Blue Hill Observatory. His Ivy League pedigree prompted plenty of jokes in the newspapers. "A Harvard man will plan New York's rain. . . . We are getting our umbrellas from Yale and our rubbers from Princeton," cracked syndicated columnist H. I. Phillips.

Schaefer and the Project Cirrus team had worked with Howell in 1948 at the Mount Washington Observatory, where he was a director. "The two scientists swapped information on the behavior of clouds and carried on some informal seeding tests from the mountain top," the *New York Herald Tribune* later reported. The pair became friends and kept up a cordial "Dear Vince . . ." / "Dear Wally . . ." correspondence for many years. In August 1949 Schaefer perfectly described Howell's virtues when he said that with cloud seeding, "the degree of success which may be anticipated in the near future shall be in direct ratio to the enthusiasm, initiative, curiosity, and perseverance of those concerned"—in addition, of course, to the necessary scientific know-how. Schaefer called Howell the day Langmuir delivered his bombshell speech to the AMS, and said simply, "You might get a call from New York City."

No doubt startled, Howell sought advice from Kenneth Spengler, executive director of the AMS in Boston. He didn't realize that Spengler was already in touch with Chief Reichelderfer of the Weather Bureau about how best to keep the city's sudden enthusiasm for rainmaking on track. "It might as well be you," Spengler told Howell. He added, "It would benefit the profession if you asked a top fee, $100 a day." A Howell associate recalled decades later that the fee "was big, big money" for a meteorologist in 1950—about equal to $1,000 a day now. At a city hall press conference on February 20 with Howell and Commissioner Carney, Mayor O'Dwyer announced that the Harvard man would conduct a preliminary study into relieving the city's water shortage through a scientific rainmaking effort. This study would lay out the requirements for such an effort, in terms of cost, personnel, equipment, and atmospheric data.

"Dr. Howell said last night he would seek Dr. Langmuir's assistance and the benefit of his research in setting up any machinery that might be decided upon for a rain-making experiment," the *New York Times* reported. "It is understood that he will not receive any pay for his preliminary studies." Howell seemed satisfied with the city's cautious approach. "It is a challenging experiment if they go through with it," he said. "But it would be nothing more than an experiment." In the hometown of General Electric, people took a broader view. "Any knowledge gained from a rainmaking study at this time may prove of inestimable value to New York for centuries to come," the always-loyal *Schenectady Gazette* declared in an editorial. "It might result in an entirely different way of obtaining dependable water supplies not only for New York but countless other cities whose sources of supply are governed by similar cloud and weather conditions."

Water levels were slowly ticking upward but overall remained so far below normal that city officials didn't dare count on the trend continuing. By the end of the month, they were anxious to get the rainmaking project moving. The reservoirs were at 46.9 percent of capacity: 106 days' supply before water pressure failed. Time might have been even shorter, as drought-stricken Cape Town, South Africa, would realize while counting down to "Day Zero" nearly seventy years later. A municipal water system won't literally go dry, *Time* magazine reported in 2018, because, "in most cases, reservoirs can't be drained to the last drop, as silt and debris make the last 10% of a dam's water unusable." If the 1950 drought continued and the worst happened, New York would be the first major city anywhere ever to run out of water.

To avoid such an unthinkable calamity, Commissioner Carney hoped to have city reservoirs full by June 1, when the summer water demand began to climb. But others remained skeptical that a rain-making project would help him meet that goal. Langmuir's old tussle with Chief Reichelderfer broke back into print, the *New York Times* noting that the city hadn't asked the Weather Bureau for help with its rain project. Reichelderfer did offer the bureau's assistance in designing equipment and evaluating results, but added a caveat: "He indicated his belief that Dr. Irving Langmuir, with whom the city's water officials consulted, has not yet proved his case for artificial production of rain." And there were those who worried that rainmaking efforts could actually have a negative effect; a Harvard meteorology professor and colleague of Howell's at Blue Hill had to reassure Bostonians that even if New York was successful in producing rain over the Catskills, "there would be no appreciable reduction in natural rainfall over New England."

Howell went down to New York for a press conference on Tuesday, February 28. He promised his preliminary report by the end of the week and said he could then begin limited operations within seven days "after we get the green light to go ahead." Carney assured the assembled reporters that he was "eager to get to it as fast as we possibly can." But science was science, and nothing moved as quickly as the O'Dwyer administration now desperately desired. Howell later admitted that it might be "six months to a year" before the project was fully underway. He also considered it "rain stimulation" rather than "rainmaking." No one could make rain, but they could perhaps stimulate more precipitation from storm clouds than would have fallen anyway.

The meteorologist completed and mailed his report to Carney as promised. Dated March 2, it stated that "the prospect that the natural rainfall and snowfall over the Catskill watersheds might be increased by a significant amount—a few inches in the course of a year—by artificial means is good enough to justify a concerted and sustained attempt to do so." Howell told a reporter by phone from his home in Massachusetts that preparing the report had convinced him that seeding clouds in an attempt to ease the water crisis was "worth undertaking."

Mayor O'Dwyer established an advisory commission on rainmaking to study Howell's report. All six members were scientists affiliated with the city's universities. Reservoir water levels, meanwhile, had fallen four days in a row. Chief Engineer Clark blamed the decline in part on tenement residents who kept taps open overnight to prevent pipes from freezing during record low temperatures. "It's a shame," he said, "that a small minority, particularly those in cold-water flats, are dissipating the great achievements of most of the people." He didn't suggest what they might do instead to prevent burst plumbing. Clark added that the cold also prevented what he called the "frozen assets" of snow and ice in the mountains from melting and refilling city reservoirs.

The venerable Brooklyn Dodgers baseball club stepped up to the plate to offer its own unique answer to the crisis. "While the winds tore across Ebbets Field in Brooklyn yesterday afternoon, a crew of workmen—headed by a Yankee rooter—began drilling a 135-foot well just outside of the foul line in right field," the *New York Times* reported. The drilling began on a Thursday dry day after a night of driving rain and lightning, while "Dem Bums" were in Vero Beach, Florida, for spring training. The Dodgers planned to use the well to water the infield and outfield grass and wash down the seats during the season. Commissioner Carney watched as a tall drilling rig rose inside the empty ballpark like a Texas oil derrick; he complimented the Dodgers on their "foresighted policy." "I think we're the first in the major leagues with a well of our own," the club's business manager said.

O'Dwyer, Howell, and the mayor's six scientific advisors gathered at city hall on the evening of March 13 and unanimously agreed that the city should proceed with the rain stimulation project. "Has there been any discussions of any legal tangles that might arise?" a journalist asked. "Couldn't imagine any," O'Dwyer said with a smile. Howell also cautioned that it would be impossible to tell on the ground whether rain that fell from any particular storm was natural or the result of the city's rainmaking activities. "We can't look for any spectacular results," he said.

New York City made the rainmaking project official the next day, when the Board of Estimate appropriated $50,000 for a six-month experiment—"the first time rainmaking has been attempted scientifically for a practical purpose." The *Brooklyn Eagle* explained that the relatively modest sum (roughly equal to $500,000 today) was "set aside for police airplanes as well as mobile trucks and radar equipment. The planes and trucks will roam the watershed area and the skies above it, looking for clouds. Radar is needed to locate clouds not easily recognizable by a ground observer's naked eye." An early offer of a plane from American Airlines was never accepted.

The city engaged Howell as a consultant for his suggested daily $100. He retained his Harvard position and would charge New York only for days actively engaged in the project; the maximum total monthly fee was $1,500. While unheard-of for a meteorologist at the time, it was hardly extravagant by metropolitan standards. Nevertheless, the round number struck the public imagination. Howell would be identified as "New York City's $100-a-day rainmaker" in hundreds, perhaps thousands, of newspaper articles during the course of the project. According to the *New Yorker*, Howell figured the contract "makes him the highest-paid cloud physicist in the world. He's hoping to replace his nine-year-old Studebaker with a new car."

Dr. Victor K. LaMer, who was both chairman of O'Dwyer's advisory committee and a chemistry professor at Columbia University, was reassuring in reply to a question from the mayor during the Board of Estimate session. "By watching for a proper time, I think there is a very good chance for success," LaMer said. "Your committee favors the experiment." Howell also thought they stood a good chance of success, and that limited operations could begin within a week. "I am sure that at times we can increase precipitation," he said, "although there'll be times when we cannot."

Newspapers anointed the project with various headline-worthy names—Operation Cloudburst, Operation Rainfall, Operation Pluvius (after Jupiter Pluvius, the Roman god of rain)—but none stuck. Officially, it was the Precipitation Stimulation Project of the City of New York. "The rain-making experiment that New York City is about to make to replenish its water supply is the most extensive and scientifically the most important that has ever been made," the *New York Times* declared. Howell's work would likely settle the disagreement between Dr. Langmuir and the US Weather Bureau. More than that, if he succeeded in making rain, the program would be "of world-wide importance, for it will be possible to overcome many droughts."

"With a wild burst of enthusiasm, the city authorities have hired a rain-maker," the *Herald Tribune* noted. The paper regretted in an editorial that New York hadn't acted faster, but hoped the scientist felt "under no immediate compulsion to rush to the mountains and turn on the celestial tap. . . . Good luck, Dr. Howell; the whole city is looking ahead to the first summer by science."

To paraphrase Winston Churchill during World War II, Howell's consultancy contract represented the end of the beginning for the city's rain campaign. With the same aggressive, confident spirit born of Allied victory in both Europe and the Pacific during the war, New York City would now begin marshaling air and ground forces, which would attack, strafe, and bombard nature's stubborn clouds until they had surrendered their rain to the parched Catskills.

Wallace E. Howell was suddenly New York City's poster boy for science. Newspaper columnists and other professional wise guys never learned that the *E* was for Egbert, and so missed an opportunity to mention

W. C. Fields and his character Egbert Sousé in the movie *The Bank Dick*. Egbert actually was Howell's mother's maiden name; friends and family called him Wally. The Boston meteorologist would never offer the sort of flamboyant gestures or flashy quotes that earn a nickname from a metropolitan copyeditor, but his meticulous nature and boy-next-door personality were appealing. "The truth is that Dr. Howell carries himself with refreshing modesty," the *Herald Tribune* stated. "His program is simple and straightforward." Even newspapers in those areas that disapproved of his rainmaking would be respectful in their coverage, only once or twice unwinding enough to call Howell "Doc" in an article or headline.

Born on September 14, 1914, in Central Valley, New York, Wally was the youngest of James and Alice Howell's four children. The couple pursued mismatched careers. James attended Columbia University and dabbled in several professions: law, investments, teaching, real estate. A uniformed YMCA secretary during World War I, he reached France only after the armistice. Alice, in contrast, focused solely on education. A Smith College graduate, she earned a master's degree from Yale University after Wally left for college. "Her thesis, as an educator, was that there is no such thing as a 'problem child,' but only the problems of children." Alice served on the state board of education in Connecticut from 1935 until 1942. During much of the 1940s, she worked at Yale as liaison between the education department in the graduate school and the pediatrics department of the medical school. Except for a short period following World War I, when both taught at a progressive school in Alabama, Alice and James reared their family largely in Connecticut.

Their youngest child followed a meandering path into rainmaking. Wally Howell graduated from Tabor Academy in Marion, Massachusetts, in 1932 and entered Harvard to major in physics. "I didn't find the theoretical side too inviting at the time," he said years later, "so I went into the applied fields—meteorology, radio communication, and the like." Howell was a wiry, athletic young man. He was coxswain of a freshman crew, swam on the varsity swimming team, and joined the Mountaineering Club. One of his professors was Charles F. Brooks,

founder of the American Meteorological Society. Brooks once asked
Howell whether he intended to continue in meteorology. The soft-
spoken student said yes, but Brooks had difficulty hearing and didn't
understand the answer. A second reply also went unheard. "For his
third attempt, Howell shouted, 'YES!' This not only convinced Brooks,
it convinced Howell as well."

After receiving his bachelor's degree in 1936, Howell entered
commercial aviation in Kansas City. He earned a pilot's license while
employed as a meteorologist by three airlines, Hanford, Mid-Continent,
and Trans World. Returning to Boston in 1939, he worked for
Boston-Maine Airways, the Yankee Network Weather Service, and the
US Weather Bureau. He also resumed his education, earning a master's
degree at MIT in 1941.

With the United States on the brink of entering World War II,
Howell was then drafted. The army commissioned him and put his
expertise to good use, although as a meteorologist rather than a pilot.
He served first with the Air Corps Ferrying Command outside Detroit
and later overseas. His climb was remarkable. "In five years he rose from
buck private to lieutenant colonel and did about everything meteorological the army thought up, from measuring rainfall to research on
sea breezes." Howell ended the war as an operational long-range weather
forecaster for the Fifteenth Air Force in Italy.

Resuming his civilian career in 1946, Howell became acting director of the Harvard–Mount Washington Icing Research Project and
a research fellow at the Blue Hill Observatory. After he and Vince
Schaefer met on Mount Washington and commenced their years-long
correspondence, the GE scientist shared his technique for making snow
in a freezer. The two men began a friendly competition to see who
could first transform a natural cloud. Schaefer succeeded over Mount
Greylock. "Shortly thereafter, Howell placed a piece of dry ice the size
of his thumb in the tower of the observatory atop Mt. Washington and
watched a passing cloud turn to ice crystals."

Howell earned his doctorate from MIT in 1948, with a dissertation
on the growth of cloud drops in uniformly cooled air. He was something

of a pioneer, since the science of weather behavior was still a young field of study. "Only since 1925 has it been possible to earn a college degree in the subject," the Gannett News Service reported as 1950 approached. "As a result, there are less than 2,500 graduate meteorologists in the country. The Weather Bureau employs about two-thirds of them, [and] could use several times that many more if they could be found."

Never a large man, Howell stood five feet nine inches tall and weighed 140 pounds. Saul Pett of the Associated Press later described him as looking "like many other young scientists you may have met in research laboratories or on university faculties. He is thin and serious and wears a crewcut streaked with gray." Had he come along a few years later, people might have remarked on his likeness to Sergeant Joe Friday, played by actor Jack Webb on the hit television show *Dragnet*. Howell's clean-cut appearance and just-the-facts-ma'am attitude matched Friday's as well. High achievers were the norm in his family. His brother was a social economist and his elder sister was affiliated with the United Nations' World Health Organization in Geneva.

When New York City called in 1950, Howell was living in Lexington, Massachusetts, outside Boston, with his schoolteacher wife, Christine, and their three young children. The area was a creative hothouse, home to architects, physicians, researchers, and professors; their neighborhood was designed by an influential, Bauhaus-inspired group known as the Architects Collaborative. Howell's son Stephen recalls growing up in a "very high-powered, intellectual, cultural environment." His father also shared his enduring love of the outdoors. Howell was a member of the Appalachian Mountain Club, but hardly had time to spare for hiking or his other hobby, photography.

Saul Pett's profile of the meteorologist went beyond physical description to report, "He is also very busy." Besides juggling his professional duties with raising a family, he was trying to finish building and painting his house and complete its landscaping. Howell kept in constant contact with New York City and the weather forecasters, and conceded that during the early stages of the city's project "there was a lot of pressure on me." As was also true for Commissioner Carney, the rain stimulation project would become the most public challenge of his career, and the one most connected with his name all the rest of his life.

"If in the long run we are able to nudge the climate in the direction of moisture over the Catskills, we will have increased the safe supply of the city's water," Howell told journalists at city hall the night before the project was funded. The venture would tell scientists how and where they could make rain in dry years, and whether doing so would help or not. "This will be more than man has known before," he said, "and he's been working at rain-making since the dawn of civilization."

6

Headquarters

To the dismay of resort owners and others, Howell planned to focus the city's rain stimulation efforts solely on the Catskills. In those remote mountains, the meteorologist later wrote, "there is less likelihood of interference with normal activities than in the thickly populated Croton and Long Island watersheds." The lack of commercial air traffic over the Catskills meant the airborne seeding operations could proceed freely, "while only a very small portion of the Croton and Long Island areas lay outside of the civil airways." Howell also expected that the mountains "would tend to intensify and localize the effects of cloud seeding, and would increase the frequency with which favorable weather situations would be encountered."

His first task was to establish a field headquarters somewhere in the Catskills. Howell began the expedition to find that site on St. Patrick's Day, three days after his hiring. He had anticipated a one-day trip, but now thought it might be two or three. Howell flew down from Boston to Albany, arriving at ten o'clock that Friday morning. He conferred at the airport with Commissioner Carney and Chief Engineer Clark, both just up from New York. Two city engineers, Fred Stein and Johan A. "John" Aalto, were in Albany as well. Stein was responsible for the water system north of New York, while Aalto had charge of the Catskill watershed.

Howell was becoming big news in New York, so ten reporters and photographers turned out, too. They included Robert Eunson, a veteran Associated Press correspondent; Richard K. Winslow of the *New York Herald Tribune*; Charles G. Bennett of the *New York Times*; and William Lowenberg Jr. of the *Albany Times-Union*. The newsmen were surprised by Howell's confidence and boyish looks. "Why shouldn't he be confident," someone cracked. "For $100 a day he ought to be able to make it rain in colors."

Albany was an ironic point from which to begin the headquarters hunt; the mayor there already had complained about New York's plans. "Mayor Erastus Corning was in a terrible swivet—he accused New York City of aiming to turn upstate New York into a desert by drying up the clouds before they got there," *Time* magazine chortled. Corning believed various "hazards" of the project should be studied and "rigidly controlled" by the state's Water Power and Control Commission. Making rain on one watershed, he feared, might seriously affect rainfall in another.

"The particular interest of the City of Albany is that a part of our water supply comes from a stream tributary to the Catskill Creek," the mayor wrote. "The Catskill watershed is immediately adjacent the Gilboa [Schoharie Reservoir] and Ashokan Watersheds of the City of New York." Corning asked whether the state commission had the authority to regulate and control artificial rainmaking. If it didn't, he wanted quick action on legislation to grant the commission the authority. The US Weather Bureau wisely stayed out of the dispute. "It's a legal problem more than a meteorological one," said the chief meteorologist in the Albany office.

Howell addressed Mayor Corning's fears while speaking to the reporters at the capital city's airport. "There's no evidence yet that the clouds we use to make rain from will be stolen from anybody," he said. ("We'll talk about that later," Carney interjected.) "It isn't likely that Albany will be affected," the meteorologist continued. "We will be operating from high ground—much higher than Albany—and for that reason alone nothing should happen to the rate of rainfall in Albany."

When the expedition set off for the Catskills, New York's water officials set up a "temporary moving headquarters in a shiny black department limousine," Winslow wrote, "and headed south to the watershed area, some of whose residents are hostile to the city plans." Bob Eunson took a more literary approach: "A boyish looking Harvard professor with a crew haircut rode adventurously into Rip Van Winkle land today in search of rain for New York City." The seven-vehicle convoy left Albany in clear, cold weather. On such a day, under a late-winter blue sky, stimulating any sort of rain or snow was difficult to imagine.

The caravan reached the town of Saugerties on the west bank of the Hudson River at noon. There it would turn right into the Catskills. It dawned on Commissioner Carney that he wasn't prepared for a snowy climb. The former typing champion stopped the limousine at a country store and bought a new pair of galoshes to cover his shoes. Later that afternoon, he would glance at his fine suit and overcoat, remember the Irish holiday he was missing in the city, and ruefully remark, "When I planned to parade today, I didn't have any intention of parading up here."

The expedition's destination was Overlook Mountain, which rose to 3,150 feet a few miles north of Woodstock. The drivers cautiously nosed up a steep, gutted road, which was covered in places by two feet of slushy snow. They made it only about halfway up before everybody had to pile out and begin hiking. Carney, Clark, and most of the party soon turned back. Athletic and accustomed to mountains, Howell kept climbing, with Stein, Aalto, and a few journalists "panting, puffing and blowing in his wake as he broke trail through the crusty snow," as Lowenberg put it. Their goal was a building the journalist called a ghost hotel, suggested by the Water Department as a potential headquarters. Built in 1929, the three-story concrete structure had stood abandoned to time and weather since the Great Depression, rising "in lonely grandeur at the edge of a cliff near the peak." A member of the expedition found a decomposing piano in an outlying cottage, Lowenberg wrote, and "pounded out a few discords."

Howell judged Overlook unsuitable while they were still on the mountainside. The site overlooked the entire Ashokan Reservoir to

the south, but a ridgeline blocked the view of the Schoharie Reservoir
to the northwest. "We can't do much here with the mountain over
there. Our radio will pick up too much ground clutter," Howell said.
Just as bad, despite its polished-granite columns and a cupola upon
which a generation of hikers had carved their names, the building was
a shambles. "We would use up all of our $50,000 fixing that place
up," he said.

The Harvard meteorologist and his hardy few headed back down the
mountainside. At the foot of the trail, somebody—likely an irreverent
reporter who had abandoned the snowy trek earlier—had pinned a note
to a tree: "Carney Slipped Here." Their day done, the commissioner and
engineer Clark motored back into New York, to enjoy what remained
of St. Patrick's Day.

———————

The great headquarters hunt resumed on Saturday morning at Kingston,
minus Carney and Clark. Howell inspected the dike that bisected the
Ashokan Reservoir, then the gauging station on Esopus Creek, before
heading west toward Slide Mountain. He would cover 150 miles today
and largely circumnavigate the rugged Catskills. Driving was perilous,
with snow still blanketing the mountains, snow squalls sweeping the
roads, temperatures remaining below freezing, and winds gusting to
thirty-five miles per hour. It was no consolation to the journalists, but
conditions were even worse out on the Atlantic for ships heading toward
New York City. "The liner *Washington*, inbound from Europe on rough
seas, reported a shipment of goldfish were doing everything but hanging
over the rail with the passengers." (The Bronx Zoo aquarium confirmed
that, yes, fish really could get seasick.)

Eunson of the AP remained laconic and unfazed. He had once
leaped overboard from a landing craft amid a hail of machine-gun fire
while covering the Pacific war. A wintry drive through the Catskills
didn't worry him. "At one point the state car driven by Aalto took a
bad skid and almost plunged from the road," he wrote. "In the climb

up Slide Mountain only two cars in the seven-car caravan could make the steep ascent in the snow." The other cars had no tire chains, and none would manage to scale any mountain the team visited today. At Slide Mountain, elevation 4,204 feet, the caravan got only as far as Winnisook Lodge. Howell pulled on a blue stocking cap, clambered onto the roof of a green frame building, looked around a bit, and came back down, not greatly impressed. Three or four feet of snow barred him from hiking higher up the mountain, but Lowenberg admired Howell's "athletic abilities that [had] left others in the party gasping in his wake the day before on Overlook Mountain."

The expedition continued westward to 3,700-foot Balsam Lake Mountain, near the village of Fleischmanns. Howell's car got stuck in a snowdrift halfway up, but he appreciated the long view to the west, where many spring and summer storms would naturally first appear. The caravan then circled back to the northeast, to the final stop in the town of Hunter at 1,900 feet. Howell inspected a facility for boys there called Camp Loyaltown, which the owners had offered for office space and equipment storage. The weary trekkers finally drove back south to Kingston, where their day had begun. Howell was noncommittal on what he had accomplished, if anything, during his two-day, three-hundred-mile expedition.

"We haven't anything decided yet," he said. "We're calling off the inspection tour for Sunday." He headed home for Boston instead. "The Harvard scientist indicated, however, that the base for the rain-making experiment might be established at a convenient spot part-way up a mountain," the *Herald Tribune* reported. "Only the high-frequency radio transmitter would be placed on the peak and used by remote control to direct the planes and trucks of the rain makers."

Catskill residents remained deeply skeptical of the whole undertaking, many believing that Mother Nature was best left unbothered. "They don't want to see people go without water in New York city, but they're afraid of what the manmade rains might do in the narrow Catskill valleys," reporter Lowenberg wrote later. A general store operator in Big Indian feared that Howell might bring down too much rain atop the spring snowpack, melting it and causing destructive floods. "Besides, this is a summer resort area," the man said. "We don't want a lot of rain to spoil the vacation season. Most of us make our living from the vacation trade."

People felt much the same in tiny Phoenicia, northwest of the Ashokan Reservoir, where many remembered a damaging flood sixteen years earlier. "Maybe they ought to try going to church and praying instead of all this rainmaking business," a local druggist suggested. In Conesville, east of the Schoharie Reservoir, a town board member bluntly asked, "Isn't the weather an act of God? And does New York City have the right to interfere with it?" Such questions would continue to dog the rainmaking project, particularly in rural areas. Devout urban Catholics, meanwhile, continued to pray for rain in the city.

———————

New York's project made other people unhappy as well, if for entirely secular reasons. Harry Grossinger, owner of the fabled Grossinger's resort near Liberty, New York, southwest of the Ashokan Reservoir, was having labor troubles with the Hudson Valley District Council of Carpenters. They charged him with unfairness to organized labor. When Grossinger offered the resort's private airport to the rainmakers, the carpenters threatened to mount a unique airborne picket line. Their president said that "two or three airplanes are being painted with appropriate signs and will be flown by members of the Liberty carpenters' union." In the end, the airport never became a contentious issue—but the sight of union planes confronting NYPD aircraft might have offered lively entertainment for Grossinger's summertime guests.

Edward Clark also dampened Howell's parade over the weekend, calling the rain project a "doubtful proposition." The Water Department's chief engineer made the off-the-cuff remarks on Sunday morning during a communion breakfast for 650 Manhattan College students and alumni engineers at the Waldorf-Astoria Hotel. Without mentioning the Harvard meteorologist by name, Clark acknowledged the conflicting fears and hopes generated by the city's project, and a public perception that he personally was skeptical of rainmaking in general. He had spoken to several scientists about it, he said, "and they just could not seem to agree as to how much rain could be made artificially, or how to control the amount or the location. They did say, however, that they could also prevent natural rain by overseeding the clouds. They have been contradictory."

Clark was right: scientific opinion *was* contradictory and not fully formed. Howell, however, saw the $50,000 project as a laudable and unprecedented experiment by a city government, an opportunity not only to help alleviate a grave civic crisis but to apply scientific rigor in collecting and analyzing massive amounts of data that would go a long way toward explaining how the natural rainmaking cycle actually worked—and how it didn't. As reporter Winslow aptly wrote in a profile for the *Herald Tribune* a week later, Howell, "but for the water shortage, might have spent the rest of his days figuring out how clouds act on paper instead of over the Catskills."

The US Weather Bureau—so often at odds with Dr. Langmuir—now helpfully held its fire on the city's plans. The *New York Post* on Friday had quoted the bureau's chief, Dr. Reichelderfer. "There've been reports that the Weather Bureau is opposed to experiments on artificial rain-making. Nothing is further from the truth," he said. "We're holding tests ourselves, in fact, and in those tests we feel we've produced small amounts of rain and snow where they wouldn't have fallen naturally. We believe, however, that the amounts were too small for the work to be undertaken on a grand scale." The Chief added that the weather bureau basically agreed with Langmuir. "We can't deny that precipitation has been produced," he said. "We disagree as to the quantity of it."

A few days later Reichelderfer praised the city's efforts, saying, "We like the way New York City is going at rain making." He especially appreciated Howell's "statistical approach." In fact, the Chief got along much better with the meteorologist than he ever had or would with Langmuir; years later, Reichelderfer would call Howell "one of the most responsible of those who are experimenting and operating in the cloud seeding field." For the city project, the Weather Bureau agreed to supply special weather forecasts through the forecast center at LaGuardia Airport, cooperate in a climatic study of the Catskill watershed, and help Howell obtain advanced radar equipment from the US Air Force.

On Sunday night, Howell spoke candidly at home about the weekend's disappointing hunt for a Catskills headquarters. "I still hope something turns up better," he said. "Our search for a site hasn't narrowed down to any two or three possible selections. None of the places we have seen so far is entirely satisfactory." The project had one bit of encouraging technical news, however. Two police department aircraft were now equipped with two-way radios; operating on 155 megacycles, they could reach from the watershed all the way to Manhattan. "It was additionally reported that the acquisition of additional police planes was being speeded to make them available for test and that radar installations might be necessary."

New York City needed every available tool and technique to meet its worsening crisis. The reservoirs on Sunday morning stood at 52.6 percent of capacity. While much better than readings at the start of the year, the figure was frighteningly far below the 92.8 percent on the same date a year earlier. No city official could seriously contemplate reducing municipal rainmaking or conservation efforts until the Catskill and Croton reservoirs held much more freshwater than they did in mid-March 1950. "At normal consumption there remains about ninety days' supply before pressure fails," the *Times* reported. The city either could back Howell's plan or find an alternative solution. But it had to take action immediately or face potential catastrophe.

7

Hurricane King

Many Catskill residents worried about New York City's rain-making efforts causing catastrophic flooding in their mountain valleys. Their concerns were understandable. After all, Dr. Howell and his team were attempting to tinker with nature, which was unpredictable enough when left alone. And observers at the time might have had particular reason to be suspicious of Howell's project if they'd read about Irving Langmuir's and Vincent Schaefer's work with hurricanes three years earlier.

General Electric signed the contract with the US government that would give rise to Project Cirrus on February 28, 1947. The document called for "research study of cloud particles and cloud modifications" and partnered the company with the US Army Signal Corps and the Office of Naval Research, supported by the Army Air Forces, soon to become the US Air Force. When GE and the Army Signal Corps announced the project's commencement shortly thereafter, the *New York Times* chirped that it was "a vast weather research program, designed primarily to disperse fog and clouds over airports, but that eventually may lead to the 'manipulation of gigantic natural forces for the benefit of mankind everywhere.'" It was an oddly cheery description of an American military program, not some humanitarian mission sponsored by the young United Nations. "If the Army could find a way to trigger precipitation processes, its forces could precipitate clouds before they could interfere with military operations, dissipate them to remove cover from enemy staging areas or targets, or precipitate clouds on enemy troops, hampering ground operations," a modern historian writes. Still, the *Times* description wasn't inaccurate, either. Humanitarian applications were evident.

The military partners controlled the flight program, while the GE Research Lab's role was "confined strictly to laboratory work and reports." The arrangement was fine with GE. It gave its researchers four-engine bombers to fly around in rather than rented single-engine cabin planes. The operation quickly became military in scale, its resources including a B-17 Flying Fortress, like those that had bombed Nazi Germany during the war, borrowed from a military weather squadron; access to other aircraft as needed; extensive flight facilities at Olmsted Field in Middletown, Pennsylvania; a weather observation station at GE's hangar; a communications network; and a large array of meteorological instruments and equipment, much of it developed by the company. Personnel included Langmuir, Schaefer, and Vonnegut, usually three or four other GE researchers, and as many as forty aircrew, weather technicians, and civilian employees. The project was up and running by autumn 1947.

GE's vaguely worded mandate covered many possibilities. Given his enthusiasm and outsize ideas, it's hardly surprising that Langmuir soon directed their work toward hurricanes. As the *New York Times* noted, a hurricane generated enough energy to power every machine on the planet for three or four years. "Yet the Army, the Navy, and the General Electric Company are collaborating in a daring meteorological experiment which is to determine whether or not the colossal vortex that we call a hurricane can be broken by making it precipitate the thousands of tons of water that it contains."

Project Cirrus hoped to seed a big hurricane churning near Florida with dry ice in mid-September 1947. "Effects of the seeding attempt to make the storm abort itself would not be immediately evident . . . and the experiment would continue for at least two days," the United Press reported. "If it works moisture carried in the tropical maelstrom would be precipitated as rain or snow, thus shattering the force of the blow." The Project Cirrus team lacked sufficient time to prepare for the September storm, but another blew in a month later. "Secrecy shrouded the plans of the government's 'hurricane busters' last night as a tropical storm gathered momentum in its drive toward Cuba and southern Florida," the *Schenectady Gazette* reported on October 11. One of the scientists (perhaps Schaefer) said the team didn't expect to stop the storm. "At this point we are interested only in seeing and recording any effects the dry-ice techniques will have," he said.

The Weather Bureau didn't yet name hurricanes. Under an alphabetical tracking system, this storm was dubbed Hurricane King; today it's sometimes called the Cape Sable Hurricane. This "small freakish hurricane" inundated Miami and southeastern Florida the next day with the worst flooding in thirty years before sweeping off into the Atlantic. "Army and Navy hurricane 'buster' planes reconnoitered the hurricane as a preliminary to seeding it with dry ice pellets in the first field test to produce precipitation," the Associated Press reported. Researchers thought the young storm would provide better data than a more mature hurricane. They didn't know what seeding it might do, but an Air Force spokesman reassured the AP, "There aren't enough planes and pilots in the United States to reduce a hurricane's potential by five percent by dropping dry ice."

According to a UP report, "The fliers waited until the storm passed across Florida and out to sea because at the present stage of scientific knowledge meddling with nature in its most violent mood might lead to unfavorable results." Another of the many wire stories from Miami added, "Even if the experiment proves successful and causes the winds to calm to normal—more than is anticipated—the hurricane will have the last ironic laugh. The experiment will have been too late this time."

The Project Cirrus B-17 began seeding at twenty-two minutes before noon on October 13, 1947, as the storm twirled like a top 350 miles east of Jacksonville. The plane dropped 80 pounds of dry ice in thirty minutes through the bomb bay doors onto the upper cloud shelf, along a track more than a hundred miles long. The bomber also made two mass drops of 50 pounds apiece. For various reasons, the researchers didn't seed the hurricane's eye or an especially active squall line. A photoreconnaissance B-17 flew 3,000 feet above and a half mile behind the lead plane. Vince Schaefer rode buckled into a third aircraft, a huge B-29 Superfortress based in Bermuda, which flew 5,000 feet above and ten to fifteen miles behind the second as a control and observation plane.

Nothing very dramatic happened at first. Schaefer later wrote that the seeding produced snow showers and light rain in areas that were above the freezing point. He didn't notice any big buildups of clouds along the seeding path. He would even question in his report whether more tests with hurricanes could be justified until "urgent and much simpler operations are completed at Schenectady." No one was concerned as the big bombers turned away toward the mainland. Nobody was watching when Hurricane King made a sharp left turn and barreled down on the Georgia coast.

HURRICANE DISSIPATES AFTER RAKING SAVANNAH, blared a front-page headline in the *Schenectady Gazette* when the GE researchers got home. The article told of flooded streets, smashed windows, lost roofs, cabin cruisers washed atop causeways, and a thousand people sheltering in the Savannah city hall. In about two hours, a moderately intense hurricane that had been behaving normally "surged to super-hurricane velocity, something previously unheard of," said Grady Norton, chief of the Weather Bureau's Storm Warning Service at Miami. Estimated damages ultimately climbed to $5 million.

The researchers were shocked, and they soon found themselves facing journalists clamoring for answers. "Frankly, we don't know just what happened and will not know until we have completed a thorough study of the data obtained by eye, camera, and instrument," Schaefer said. He added that "everything clicked perfectly and we believe we have a great deal of important data." He wasn't allowed to say much else; results of the mission were classified, despite earlier army assurances that no secrecy would be attached to the tests.

Had Project Cirrus, in fact, caused Hurricane King to veer sharply ashore into Savannah? Schaefer didn't think so, and repeatedly said as much. The Weather Bureau also doubted it. "Only the Army, Navy and scientists co-operating with them will be able to say what the dry icing did to the hurricane. Had there been no such experiment, the storm developments may have been the same," Norton from the Storm Warning Service said.

Dr. C. G. Suits, the GE Research Lab director, insisted nearly a year later that Project Cirrus's storm experiments had been misinterpreted. "At no time was any consideration given to attempting to control—or 'bust'—a hurricane, as was erroneously reported," he said. It was a revisionist take; a navy spokesman had clearly told reporters before the Hurricane King test, "The result of this will lead us to further accomplishments in efforts to break up a hurricane." Dr. Suits further deferred responsibility by noting that while controlling the weather might one day be possible to some degree, breaking a hurricane constituted "what you might call post-graduate weather control. What we are trying to do now is to get through First and Second grade." He added, "We have no plans to send planes into the hurricane area" in 1948.

Irving Langmuir privately believed that Project Cirrus had indeed affected the hurricane, but as a government contractor he couldn't and didn't say so at the time. It wasn't until April 1955—three years after Project Cirrus had ended—that he publicly said what he believed. "There was not one chance in seven thousand that the hurricane would have turned if it had not been seeded," the scientist said at an international symposium in Albuquerque. "I had suggested that we wait until the storm was further north and out to sea to experiment."

Langmuir later said that fear had halted the hurricane tests, and that such experiments "could lead to modification of big storms before they reached populated areas." He dismissed Weather Bureau doubts that it was possible to modify anything so immense as a hurricane. "I suppose one could not modify a forest with a match, either," Langmuir snapped. He recommended further experiments but on typhoons in sparsely populated areas of the Pacific Ocean.

This was all too much for the Weather Bureau, which was far more cautious and methodical in drawing conclusions than the dynamic

Nobel laureate. Federal meteorologists set out to prove or rebut Langmuir's contentions. They spent a year poring over weather maps and ships' logs to reconstruct the storm's path. They finally concluded that Hurricane King had been spinning in the same spot over the Atlantic Ocean between seven thirty that morning and when Project Cirrus seeded it at around noon. "Therefore there is every indication that the hurricane was standing still or already starting to recurve at the time it was seeded," the bureau announced in August 1956, "and, if it had been standing still, the chances are that it would have started to recurve soon anyway, because a hurricane usually curves in one direction or another after being stalled for a time."

The conclusion appeared to rule out Langmuir's belief that seeding had contributed to the storm's unexpected swerve. Although the Weather Bureau didn't say so, Langmuir himself had identified a 1906 hurricane that had followed a nearly identical path to Hurricane King's. The GE scientist, however, batted away the federal rebuke. "This is the kind of thing the Weather Bureau has done consistently," he retorted. "They start out to prove something and they always find something to support their contentions." He again called for a resumption of hurricane-seeding tests. "Maybe that will settle the issue."

But that dustup was several years ahead. In 1950 concerned parties still wondered whether Howell's rainmaking operations might cause a thunderstorm to slam off in some unexpected direction.

8

The Goose

Wallace Howell returned to New York City early on Monday, March 20, 1950. Spring would arrive that day at twenty-four minutes before midnight. Howell met Mayor O'Dwyer, Commissioner Carney, and Chief Engineer Clark at Floyd Bennett Field along Jamaica Bay. There the men inspected the amphibious planes the rainmakers would use to seed clouds above the Catskills. Howell told the mayor that initial seeding with dry ice would be "simplicity itself." The meteorologist was ready to go whenever the Northeast Weather Service, a private company in Boston, told him that clouds were rolling in, either from a nor'easter or up from the Ohio Valley. "As soon as they warn me, I'll rush down from Boston to Floyd Bennett and pick up one of the Police Dept. planes for a seeding run in the Catskills," Howell had told the *New York Post*. "Of course, our problem will be to seed the clouds at the right time and to make sure they lose their rain where we want it to fall—right over the watersheds." Seeding with silver iodide would commence later, whenever the equipment was ready.

Responsibility for the rain stimulation project fell mainly to Carney, and today was one of the few times that Howell personally briefed the mayor. Press photographers captured the dapper politician and slim scientist chatting beside a police aircraft. The meteorologist also briefed the NYPD pilots on seeding techniques. "We are

offering everything we can to see that science has a fair opportunity
to show what it can do on rainmaking," O'Dwyer assured the milling
journalists.

Howell appeared on local television at 8:30 that night. His segment
on WOR-TV had been scheduled for St. Patrick's Day, but the scientist
had postponed it, too busy up in the Catskills. Although still a relatively new medium, television resembled "a fast-growing child." The
number of sets in the United States was expected to reach six million
in 1950, up from three million the previous year and just five thousand in 1946. Howell's fifteen-minute program, *Scientific Rain: What
It Means*, appeared in the day's *New York Times* broadcast listings. He
went before the cameras armed with charts and diagrams to explain
the basics of rainmaking to curious New Yorkers. In print and over

(Left to right) Dr. Howell, Mayor O'Dwyer, Commissioner Carney, copilot Gustav
Crawford, and Chief Engineer Clark confer beneath the wing of the NYPD's
Grumman Goose, with three uniformed officers looking on. *AP Photo / Herb Schwartz*

the airways, the hot spotlight of public scrutiny had turned his way. "This is certainly a jet-propelled operation," he would say a few days later after also appearing on WPIX-TV. "I didn't know I was going to be a celebrity, and I don't want to be one."

———————

Tuesday was the first full day of spring but hardly seemed it, "so full of rain and snow that many drought-conscious New Yorkers suspected a trick." Dozens of people called the Weather Bureau, demanding to know whether this was "the real stuff" or the city's first artificial precipitation, although seeding hadn't even begun. Carney announced during a press conference that every Catskill site the city's team had inspected over the weekend had been rejected. Howell would look farther out, near tiny Downsville on the east branch of the Delaware River. The *Herald Tribune* described the area as commanding "wide vistas to the west, whence most of the favorable clouds will come."

The rain stimulation project made better progress at Floyd Bennett Field. There the aerial seeding team was readying for flights as soon as conditions over the watershed were favorable. The city had reason for optimism beyond Vince Schaefer's General Electric experiments. Seeding with dry ice apparently had worked well in Great Britain the previous August. A serious drought in northern England had threatened to shut down some industries in southern Scotland. In a joint project by the Royal Air Force, Imperial Chemical Industries, and a government meteorological forecasting station, an RAF bomber flew a seeding run, similar to the B-17 missions of Project Cirrus. "On Aug. 3, at 5:20 P.M., 200 pounds of carbon dioxide dropped onto cumulus clouds 10,000 feet above and to the west of the Pennine Mountains brought 'moderate' rain," the *New York Times* reported. A second drop of 300 pounds was equally successful. New York City hoped for similar results in the Catskills.

Carney said the team at the airfield was "prepared to move in on the first opportunity that becomes evident to Dr. Howell." He added, however, "the main source of [water] saving still lies with the city's residents." The rainmakers got some good news when the US Air Force announced that it was granting the city's request for radar sets. The Weather Bureau had supported this request; Chief Reichelderfer explained that radar could "detect and record more accurately than human beings when showers begin to fall," which in turn was useful in determining whether rain began before or after seedings.

Howell and Carney were eager to begin the rainmaking experiments. "But atmospheric conditions will be the determining factor, and it might be weeks before the chosen clouds are pelted with dry ice or silver iodide to produce rain for the water-short metropolis," Bob Eunson had written during the Catskills tour. The weather still wasn't right for the Harvard scientist, and the clock was ticking as Howell's little air force stayed grounded through the weekend. Columnists and reporters struggled to keep readers informed and entertained during what became the silly season.

"A scheduled rain-making experiment was postponed today because of—rain," the United Press reported on March 23. "A nightlong rain followed by driving mist led city authorities to conclude that an attempt to sow rain today would be akin to sending coals to Newcastle." Radio commentators, too, delighted in announcing the postponement of rain-making tests due to heavy rain. But the *Schenectady Gazette*, the home-town newspaper for General Electric researchers, reacted stiffly to such humor. It was no wonder, proclaimed a *Gazette* editorial, that many a curious person was willing to study rainmaking alone or for a company, "but refuses to go into any rain making venture in which he exposes himself to public responsibility. He knows he may receive little for his pains but ridicule, whether he succeeds or fails."

The *Gazette*'s lecture had no discernable effect on the jokers and nay-sayers. Writer Robert C. Ruark disparaged what he called the "voodoo of a fairly base grade" that New York City was now calling upon during the terrifying era of the H-bomb. "Also there are too many magicians in the act, including the legislators who tried to crowd through a couple of bills restricting the right to make rain. . . . Also too many props.

Airplanes, trucks, two-way radio, blowers, seeders, sprayers—this is merely 20th century complication of simple sorcery."

Columnist H. I. Phillips wrote that New York City had gone into the business of "strip-teasing a cloud. The money is up and the people hired. Dr. Wallace E. Howell . . . has signed up as Big Chief Rain-In-The-Right-Spot. He gets $100 a day. (That is more than Al Jolson used to get for his rain numbers.)" And in Texas, an itinerant Baptist preacher wrote a letter to New York governor Thomas Dewey, stating that he had prayed for blessed rain to fall on the big northern city. "Now if this comes to pass . . . ask the officials of New York to give me an offering of $7,000 to help me in my work."

David C. Whitney of the United Press imagined eavesdropping on plane-to-plane radio transmissions when what he dubbed R-day—"rain day, that is"—finally arrived.

> Howell: "Cumulus cloud sighted at 4 o'clock. Prepare for
> bombing run."
> Pilot of 2nd plane: "Roger. Pellets ready. Dry ice chutes
> warmed up."
> Howell: "You're over the target now. Ice bombs away."
> Pilot of 2nd plane: "Sighted cloud, sank same."

On Saturday, March 25, the Pure Carbonic Company delivered a rush order of 600 pounds of dry ice to the police hangar at Floyd Bennett Field. The team put it into a deep freeze, ready to be ground up and loaded onto the planes. "It must be used within a week, however, because otherwise it will evaporate." Three days later, conditions were still iffy for attempting the first seeding mission: rain and fog, with wind gusting to forty miles per hour. While exactly what New Yorkers needed to replenish the reservoirs, the foul weather also snarled traffic and delayed departures from airports and harbors. Several upstate communities reported flooding. A steamer collided with the lightship *Ambrose* in the pea soup shrouding the Atlantic. All the while, the airfield hunkered beside Jamaica Bay like a stray cat under the deluge.

New York City had established Floyd Bennett Field as its first municipal airport in 1931. It was named for a Brooklyn aviator who had received the Medal of Honor for daring flights into the Arctic with Richard E. Byrd (although whether the pair actually reached the North Pole was debatable). In 1941 the city sold the airfield to the US Navy, which used it as a naval air station during the war. The NYPD aviation bureau now occupied a navy hangar. The city had formed the bureau in 1929 to police "the new menace of modern civilization, the reckless and incompetent flyer."

By 1950 the metropolis had a dozen airports, eighteen seaplane bases, and one helicopter port, making Floyd Bennett Field no longer vital to New York City aviation. Police fliers based there now flew mainly patrol and rescue missions (although they had, in fact, cited a Northeast Airlines pilot for flying a DC-3 too low the previous July). The fliers in blue also had once tracked a homing pigeon carrying ransom to a kidnapper, resulting in an arrest. Occasionally, too, they fulfilled their original purpose of chasing "ex-GIs with combat habits, who are inclined now and then to buzz the girl friend's house in Bensonhurst or fly upside down, for the laughs, along Mosholu Parkway." The bureau's seven pilots had all served during the war themselves and were familiar with such high jinks.

Though the weather was questionable on the morning of Tuesday, March 28, plans went forward for the city's first rainmaking flight. Print, newsreel, and radio reporters waited impatiently in the hangar for Howell to arrive, and when he walked in at seven o'clock, everyone "rushed about getting final preflight interviews and pictures." Photographers clicked away as Commissioner Carney and Chief Engineer Clark watched a pilot and a mechanic load a box of crushed dry ice into the plane. Howell planned to scatter 100 pounds of the stuff on this initial mission. "You can dump dry ice into the cloud with a kitchen cup from a cardboard carton and it will work. But we will use little chutes," the meteorologist explained. He also tried to assure worried upstate farmers that his seeding posed little risk. "We have no intention of flooding people out," he said. "Floods are caused by a series of storms, rarely by one storm."

The tension, the driving rain, and the sights and sounds of a busy airfield may have reminded Howell of his wartime service with the Fifteenth Air Force in Italy. Vince Schaefer's snowmaking sortie in 1946 had sparked memories from a former B-17 pilot in the same outfit: "Many a morning, I recall being awakened before dawn, rushed to a briefing hut for instructions concerning a mission which should have been flown at once against an enemy aircraft factory, an oil field, or other vital installations, only to have the entire attack canceled because of cloud formations which hampered mass formation flying." A military historian writes, "The Fifteenth's chief adversary was not the Luftwaffe or any of the other five Axis air forces. It was the weather." Howell was now fighting the weather again, with simpler, less deadly, and perhaps more effective weapons.

A Grumman Goose was the pride of the city's air fleet and the plane the rainmakers planned on flying this day. It was a twin-engine, high-winged amphibious aircraft wearing the same green, white, and black livery as New York City patrol cars. Grumman Aircraft had named its amphibious planes in descending order of size: Mallard, Goose, and Widgeon. The company introduced the Goose during peacetime as an air yacht for millionaires. British publishing magnate Lord Beaverbrook was among the first ten people to buy one in 1936. No stranger to comfort himself, Mayor O'Dwyer loved flying on the city's reliable eight-person model, which was equipped with tables and reclining chairs. Thanks to its two 450-horsepower engines, the plane had a cruising speed of 145 miles per hour. You can see the actual aircraft at Floyd Bennett Field today, fully restored, in a hangar operated by the National Park Service.

Some observers consider a Goose homely, "like a cartoon whale balancing a surfboard on its back." Others see it as beautiful, with "an almost architectural design to the main cockpit windows that wouldn't be out of place in a gothic cathedral." The plane's unusual profile had been familiar to troops from Trinidad to Oahu during the war. American, British, and Canadian armed forces, among others, all had flown some variant of the Goose. The US Coast Guard still flew them in 1950, and the French later used them as bombers in their war in Indochina.

US Navy amphibious utility transport, a military version of the civilian Grumman Goose, flying above Alaska during World War II. *Courtesy of Naval History and Heritage Command*

The NYPD had bought its Goose as surplus in 1947 for $14,000 from the War Assets Administration. Its current value was $75,000. The aircraft's official police designation was Aviation One. The department also operated a single-engine Stinson Reliant (Aviation Two), a surplus Grumman Widgeon (Aviation Three) and a two-place Bell D47 helicopter with its distinctive bubble cockpit (Aviation Four). Only the Goose and the Widgeon were equipped to seed clouds with dry ice— the first through an apparatus fitted to an emergency door, the second through a hopper placed in a window. Dry ice was the only weapon either plane took aloft. "City planes carry no heavier armament than a cop's .38-caliber revolver," the *New York Times* wryly noted later.

Continuing rain at Floyd Bennett Field delayed the planned eight o'clock takeoff. The Goose carried fuel for three hours, the *Brooklyn Eagle* reported, "and the plan was to land it, when the fuel was near exhaustion, at Scranton, Pa., or Binghamton [NY], for refueling before the return to Brooklyn." This plan changed several times during the morning. Howell finally decided to fly toward Liberty in northern Sullivan County, then radio down to a pair of Water Department employees on the ground who would observe the clouds from a Park Department station wagon. "The rainmakers made it clear they wanted no clouds running amok when tickled," an upstate newspaper added. "All they wanted was a gentle flow of girlish tears."

Bob Eunson of the AP was again reporting on Howell's activities. Eunson waited with a science writer, a photographer, and a pilot at Flushing Airport in Queens. They expected to depart in a private plane at the same time as the Goose, then rendezvous with Howell at Liberty. Eunson watched the rain fall for hours, receiving periodic reports on why the expedition was delayed.

They were:
Lack of clouds.
Too many clouds.
Too warm.
Too cold.
Too wet—and this you can say again.

Howell checked forecasts all morning, then dashed to LaGuardia Airport for a personal update. Conditions still didn't look good, he said on his return, but cumulus clouds building over the watershed behind the passing front were perhaps ripe for seeding. The ground crew rolled out the Goose as the storm finally swept past Floyd Bennett Field at eleven thirty. Climbing aboard with Howell were his pilot, Acting Sergeant Gerald J. Crosson; his boss and copilot, Acting Captain Gustav Crawford; Water Commissioner Carney; and Dr. Carl W. Van der Merwe, a New York University physicist and member of Mayor O'Dwyer's science advisory committee.

"It doesn't look as though we'll get very good results," Howell admitted. "But we feel that it's worth a try." He had a first-rate pilot at the controls of the Goose. Crosson had been a highly decorated lieutenant colonel during the war, flying much bigger B-26 Marauder and B-29 Superfortress bombers.

The Goose lifted off at three minutes past noon. In Flushing, the AP's plane splashed through puddles to get airborne at almost the same moment. Howell wasn't 5,000 feet into the air, however, before he realized his mission was a failure. "We were held to such a late start from the airport that the clouds with which we wanted to work over the watershed had already passed too far to the east," he said. "It was unusually warm for this time of year at 10,000 feet. If the clouds had been there, we would have had to rise well above 10,000 feet to do the seeding." He could have chased the clouds and seeded them over Connecticut, he added, "but the results would have been way east of the reservoirs."

Howell stayed airborne anyway, taking the opportunity for a shakedown flight. The Grumman Goose had a reputation as "a whole lot of airplane" and required two people to fly it. Crosson and Crawford flew northwest to near Scranton, then northeast over the watershed, before heading south again for Brooklyn. The only seeding was a little dry ice tossed out by Captain Crawford, to ensure that it could clear the plane's stabilizers. "At latest reports no one had been reported injured by the ice chunks," a newspaper quipped on Long Island. The air team also tested the high-frequency radio link with the Park Department station wagon.

The sun was shining when the Goose touched down at the airfield at two fifteen, having covered 350 air miles in a little over two hours. "We didn't see a good healthy cloud all the way out," Carney told the waiting reporters. He added that the air and ground crews nonetheless had benefited from the practice. "What we learned will make it easier for us to get going next time," Howell said. "I am sure we will be able to do a better job."

The Associated Press plane had meanwhile wandered amid bumpy clouds for two and a half hours, with no news of Howell or the Goose's whereabouts. Running low on fuel, the journalists finally had touched down in tiny Moosic, Pennsylvania, outside Scranton. When Bob Eunson called in their location, an AP staffer at the newsroom back in New York didn't believe him. "You just don't talk into a telephone," Eunson wrote, "and get somebody from a place called Moosic."

Freed of fog and rain, New York City grew unseasonably warm, soon topping 65 degrees in Manhattan. The afternoon became so pleasant that the *New York Times* the next day ran a photo of three young men lounging in their shirtsleeves in Madison Square Park. Newspapers all across the Empire State noted the unexpected touch of spring and poked fun at Howell's thwarted expedition. CATSKILL SKIES TOO BLUE, THEY COULDN'T FIND A CLOUD, the *Democrat and Chronicle* proclaimed in Rochester. CLOUDS ARE TOO SHY FOR N.Y. RAINMAKERS, headlined the *Tonawanda Evening News*.

"Ironically, rainfall since the rain-making project began has been well above normal. The reservoir levels rose by 2,275,000,000 gallons yesterday," the *Syracuse Post-Standard* noted. Eunson concluded his own self-deprecating article with a plea: "Don't worry about the water shortage, doc, just wring me out." Dr. Howell was characteristically unfazed. "We got a wonderful view of the watershed," he said.

9

Who Owns the Clouds?

City hall knew that if rain stimulation succeeded, any precipitation the project produced also would create a legal quagmire. The city was dealing with both Mother Nature and human nature. Add jurisprudence, and you had a truly volatile mixture. Historically, a weather disaster had always been considered an act of God. "Once man takes control, he will be held responsible for his mistakes," British scientist and inventor Professor A. M. Low had predicted in 1949.

Vince Schaefer had realized the risk almost as soon as his veils of snow drifted down above Mount Greylock. "The snowflakes all evaporated in dry air before they touched the ground," *Life* magazine reported. "But Schaefer suddenly had a qualm. If his snow had landed, could he be sued?" General Electric certainly thought that he could be. The company's top legal minds looked into the matter and foresaw "a very worrisome hazard in this new form of cloud experimentation." The company couldn't balance the risk to its shareholders against any known benefit from new products or business, so "there was great reluctance to incur risks of uncertain but potentially great magnitude." GE Research Lab director C. G. Suits later admitted that the company "has disassociated itself legally from rainmaking in the fanciest ways it can think of."

General Electric continued to support the groundbreaking work by Langmuir, Schaefer, and Vonnegut through Project Cirrus, but because it was a federal contract, GE was protected if one of their weather projects went awry—such as a wayward hurricane barreling toward Georgia. New York City couldn't offer any similar umbrella of legal protection, which is one of the reasons GE directed the city to Dr. Howell instead of offering the company's own weather modification services.

But that meant the same murky but ominous legal issues now loomed over both Howell and the city. The Associated Press had pointed out two of the most important ones in February 1950, when Commissioner Carney and Chief Engineer Clark first visited GE in Schenectady: Did New York have a legal right to make rain? And what happened if it made too much rain in the wrong place? "Operating in the same touchy area where Irving Langmuir started," *Time* magazine later reported, "New York's Dr. Howell was warned that if he talked too much about dumping rain in the watersheds New York City might be sued by outraged farmers and resort owners in the Catskills."

It was also possible, despite assurances to the contrary, that stimulating precipitation over the Empire State "would rub out a lot of rain destined normally for Boston, and maybe Boston wouldn't like that," Dr. Langmuir warned. "Besides, producing heavy rain artificially is, in a sense, modifying the planet. Some people will be pleased but a lot of others will start hollering and there will be lawsuits." Langmuir was careful to distance himself from the city's efforts. "I am only interested in telling Mr. Carney how we have produced rain at various times," he said before meeting with the commissioner. "Any other problems will have to be solved by him." The scientist later added, "In New York City, we see only law suits, with no financial benefit."

———

Upstate resistance to the city's rainmaking plans made a point of mentioning the city's potential legal liability. Two small newspapers, published by different entities, had printed nearly identical front-page editorials raising legal concerns. "Before tinkering with nature in order to produce more water for New Yorkers there are possible hazards to this widespread region that should be considered," the *Kingston Daily Freeman* cautioned on February 20, 1950. "Of course it first must be established if New York has the right to tamper with the clouds." Two days later, the weekly *Catskill Mountain News*, in Margaretville on the opposite side of the Catskills, likewise pointed out that the area's main commercial enterprises were resorts and farms. "What would happen to these businesses if this man-made rain came down in such amounts that it flooded and otherwise damaged properties in this region?" its editorial demanded to know. "Would the city of New York pay for the loss of business and for the destruction of property?"

The *Schenectady Gazette* found no legal precedent for New York City's rainmaking plans, going as far back as old English or even Roman law. "The law department in effect would be starting from scratch . . . since no one to our knowledge has even tried to test in a court the legal responsibility of a municipality for its artificial production of real rain—or its prevention of rain," the paper noted in an editorial. Commissioner Carney himself acknowledged the issue. "We're looking into the legal side of it," he said. "Maybe we would have to get permission from the state or the federal government. We aren't sure what our responsibility would be to anyone who didn't want rain, and we don't even know who owns the clouds."

Ownership was a key and timely issue. Who *did* own the clouds, if anyone? *Time* magazine called it "a horrendous lawyer's question." The *New York Times* later asked in an editorial, "Is there any valid analogy with flowing water or right to capture roaming wild animals? Can it be held that clouds move in interstate commerce because of the economic effect of drought or precipitation on our economy?" Dr. Howell, a central figure in the debate, would write in 1951, "Common law provides no precedent that applies to weather modification; not even the doctrine of riparian rights can be applied, for atmospheric rivers do not flow between confining banks." As a *Buffalo Evening News* editorial noted, "The problem of water diversion from rivers and underground sources has long been a cause for litigation, between individuals and various

states; but the possibilities are limitless if the people in general begin
fighting over who gets what cloud."

Two California law journals had been among the first to address
the issue. An unsigned article appeared in the *Stanford Law Review* in
November 1948. "The article goes clear back to Justinian for hints of
'Who Owns a Cloud?'" the United Press reported. "The rub is that
nobody in particular owns a cloud, but everybody in general does,
because of a cloud's 'vague and elusive nature.'" The Stanford piece
concluded by stating that it would be hard to predict how courts would
rule. Those in the East likely would apply common-law rules, while in
the West individual statues would apply. "In most jurisdictions, however, the landowner stands an excellent chance of establishing a right
to water in clouds."

A sharp third-year law student at Stanford's rival, the University
of California, Berkeley, joined the discussion the following March. "It
would be nonsense to talk of private ownership of a particular cloud,
even were it a better defined and less ephemeral thing than it is," Stanley
Brooks argued in the *California Law Review*. But actually producing
rain or snow was another matter, regardless of who owned the cloud.
"There is an urgent need for clarification of the resulting legal relations,
especially in the private controversies which are likely to arise," Brooks
wrote.

The 1950 drought in the Northeast gave new urgency to the issue.
"The legal angles of weather control are already being argued out by
professors of the country's law schools," *Changing Times* reported, adding
that "most lawyers and scientists believe that rain making must be a
monopoly of the federal government."

An attorney from New York, Albert P. Blaustein, considered the
various legal angles in the *New York Law Journal*. He wrote that a "new
field of law has been created and a new set of judicial rules must be
formulated." He speculated elsewhere that complex rainmaking disputes
might lead to the formation of state and interstate "cloud commissions."
Blaustein doubted that a property owner's rights stretched as high as the
clouds, but nobody really knew for sure. It might be perfectly legal for

someone to seed clouds over your land. Equally, that person might be liable for any damage to your property caused by stimulated rainfall. "All we can do now about such matters is surmise," Blaustein said. "I can tell you, though, that when the rains come, the lawsuits are sure to follow."

———————

The first lawsuit hit Gotham before Howell even attempted to stimulate rain over the mountains. Attorney John E. Egan of Kingston filed a suit in Ulster County on March 21, 1950, a preemptive strike from the Catskills. Egan represented groups as well as individuals: a couple who owned a hotel and cottages in Shandaken; a pair of dairy farmers in the Town of Rochester; the Big Indian–Oliverea Board of Trade; and the Pine Hill and Phoenicia Chambers of Commerce. "The trade groups claimed more than a hundred representatives a piece," the *New York Herald Tribune* reported. "The total population of the four towns is less than 1,000."

The plaintiffs sought an injunction to stop New York City from proceeding with the rainmaking project. Egan charged that officials had knowingly perpetrated a hoax—a "huge and false propaganda campaign through the means of the press, radio, motion pictures, and other means, designed to create a false fear in the public mind that there existed a dangerous water shortage . . . when in truth and fact no such dangerous water shortage existed." The whole thing, Egan alleged, was a conspiracy to sell water meters valued at more than $300,000. The Water Department, in fact, had proposed a law in 1948 requiring the installation of meters in 150,000 multiple dwellings. Opposition from real estate groups had shelved the plan, which Carney now wanted to revive. Egan certainly knew about the debate and controversy. He was no random small-town lawyer but a former head counsel in the Kingston office of the New York City Board of Water Supply's claims department.

Back in the city, a smiling deputy sheriff handed summonses to Clark, Carney, and Mayor O'Dwyer on March 22. The lawsuit also named Howell, a water engineer, and three members of the Board of

Water Supply, which operated under the Department of Water Supply, Gas and Electricity. The Harvard meteorologist was away when his summons arrived, but he was not facing any personal financial liability; his contract stipulated that the city would pay any judgments against him. Still, Howell must have shaken his head on reading the complaint, which alleged a "grave danger" of his rainmaking causing floods in the watersheds, leading to "the drowning of many persons and destruction of property of the plaintiffs." The florid language and the David-versus-Goliath nature of the challenge raised eyebrows across the state. It was unclear, however, whether Egan actually had a stone in his slingshot.

"Somebody doesn't want it to rain, I take it," Mayor O'Dwyer quipped. The *Schenectady Gazette* was typically sympathetic toward the rainmakers. "At the risk of being facetious," its editorial writers opined, "if we were in a jurist's shoes and thought we could get away with it we would rule that the complainants had nothing to complain about until rain making had been attempted, after which they could try to show damage had been done." The *Brooklyn Eagle* snickered at what it called a "legal waterspout," reporting that "four fear-shaken communities up the Hudson River clouded up and rained legal charges all over the administration of Mayor O'Dwyer. . . . The city, however, plans to go ahead with its experiment just as if nobody cared."

New York had twenty days to respond to Egan's request for an injunction, which gave the city plenty of time to maneuver. "In the absence of a restraining order, we can continue with plans for the project," said Corporation Counsel John P. McGrath, city hall's top lawyer. Motions and countermotions pushed proceedings far into the summer, by which time cloud seeding had already begun and an injunction was pointless.

Egan also failed to gain traction by dredging up the water meter controversy. A New York City conservation measure would go into effect in June that resolved the tabled issue in the Water Department's favor. As the *Long Island Star-Journal* reported, the new law provided that "Water Czar Stephen J. Carney can force the installation of water meters in any building that he feels isn't complying with his water regulations." The measure didn't grant Commissioner Carney all of the authority he had wanted, but near enough.

Rainmaking opponents faced another legal setback on May 11, when a New York State Supreme Court justice sided with the city and refused to grant a temporary injunction sought by the owner of the six-hundred-guest Nevele Country Club resort in little Wawarsing, twenty miles southwest of the Ashokan Reservoir. (In New York, the Supreme Court is a trial court, not the state's top court.) In that case, *Slutsky v. City of New York*, the plaintiff claimed his investment might be jeopardized by city rainmaking. "Look, our guests come from New York City," said the country club's owner, Julius Slutsky. "They don't know much about the country. They say, 'They got rainmakers up there, so why should I go to Slutsky's?'" The court weighed potential inconvenience to Slutsky and a relatively small number of guests against the challenge of supplying freshwater to millions of people living in and near the metropolis. "The relief which plaintiff asks is opposed to the general welfare and public good and the dangers which they apprehend are purely speculative," Justice Ferdinand Pecora ruled. "This court will not protect a possible private injury at the expense of a positive public advantage."

The state government also refused to rain on New York City's parade. Attorney General Nathaniel Goldstein had said in March that the State Water Power and Control Commission had no control over rainmaking experiments, although he added that such regulation was perhaps desirable. In a legal opinion requested by the commission, Goldstein had said it was "up to the Legislature" to decide which state agency, if any, should have such unusual power. Two weeks later, as attorney Egan pointed accusingly from Kingston toward Manhattan, the legislature considered two bills supported by Albany's Mayor Erastus Corning. One would have created an eleven-member commission "to make a comprehensive study of artificial rainmaking." The other would have given the Water Power and Control Commission jurisdiction over rainmaking. Lawmakers killed both measures.

No matter what state or federal laws did or didn't yet say, ambitious rainmakers were busy far beyond the Catskills. In many parts of the country, *Changing Times* later reported, people couldn't be sure whether a local downpour was due to "the usual operations of Mother Nature, to the highly secret operations of the U.S. government rain makers, or even to the casual operations of some privately hired aviator throwing dry ice out of his plane." Legal events in New York would affect the rest of the country. Opposition had now been frustrated in both the assembly and the state courts. For the moment, at least, about the only force that could thwart New York City's ambitious plan for modifying the weather was Mother Nature herself.

10

Achilles' Heel

Thanks largely to the storm that had drenched the Catskills on the morning of Tuesday, March 28, then fled over Connecticut before Dr. Howell could seed it, more than 7 billion gallons of freshwater had flowed into city reservoirs. The rise was the largest in fifteen months, the twenty-first consecutive daily gain, and the eighth straight gain of more than 1 billion gallons. Snowmelt also contributed. "The high temperature, coupled with the rain yesterday, loosened up the frozen assets and turned them into liquid assets," said Chief Engineer Clark.

The reservoirs reached nearly 60 percent of capacity. This was well below the 94 percent level recorded a year earlier, but even so, just as the rain stimulation project was ready to go, New York City's water situation had improved. The irony of having the city's first rainmaking mission scuttled by a rainstorm wasn't lost on unhappy upstate communities. "Despite the fact the Harvard scientists are applying their test tube findings to the actual conditions, they have not been able to predict even when they'll go out to attempt to make rain," observed the *Catskill Mountain News*. "Whether, once they're actually over the 'right' clouds, they can produce man-made rain, is another question we leave for posterity."

General Electric's Vince Schaefer, meanwhile, was visiting Oxford, England, to address a meeting of the Royal Meteorological Society.

There he said that the publicity surrounding New York City's rain-making experiment had grown too sensational. "No one is a miracle man," he cautioned. "Rain can only be precipitated from a very thick cloud which in temperate latitudes like Northeastern America is not far from natural precipitation anyway." Schaefer praised his friend Wally Howell as head of the project, but added that the city would be "lucky if it gets an inch or two more of rain than it normally would." Howell likely would have agreed—or at least not loudly disagreed.

Thursday, March 30, was a water holiday. The Harvard meteorologist said he hoped for "a Dry Thursday in the city and a wet one over the Catskills." Howell didn't regret his earlier failed seeding attempt. "You can't expect rain from every flight," he said. "In fact, one of the important things we are trying to learn is when we can make rain and when we can't, so even a so-called unsuccessful flight will add to our knowledge." Thursday proved another disappointment, however: the skies stayed clear and New Yorkers performed poorly at conserving water. Howell left the city with a radar technician to renew his ground search for a Catskills headquarters and communications center. Commissioner Carney considered the Dry Thursday "a complete failure." In one bit of good news, the city announced that its Department of Marine and Aviation had captured 30,000 gallons of rainwater since January from downspouts at Pier A on the Battery in Lower Manhattan. City workers used the runoff to wash department vehicles.

Howell was still touring the Catskills the next day when the rain stimulation project received welcome news—not the right sort of clouds for seeding but more, bigger, and better aircraft with which to seed the clouds once they arrived over the watershed. Electronics company Sperry Corporation, headquartered at 30 Rockefeller Plaza, said it would supply New York City with three radar-equipped DC-3s and scientific and engineering expertise. The planes came equipped with both standard and experimental gear, including a Zero Reader automatic flight indicator, automatic pilot, and automatic approach control. Sperry attached no cost or strings to its offer and added that it would work out exactly how the city would use the planes in conference with Howell and his team. Commissioner Carney immediately accepted "with deepest gratitude." The offer "materially broadened the potential scope of the rain-making effort, especially the long-range project," Charles Bennett of the *New York Times* reported. "It ended the possibility of further delay in obtaining radar equipment and multiplied by two or three the power of the air arm of the cloud-seeking battalions."

The three aircraft, test beds for Sperry, were conveniently based at MacArthur Airport near Sayville in central Long Island. "The Sperry planes, which are faster than the police planes, can attain altitudes close to 20,000 feet, some 5,000 feet above what the city's Grumman Goose can climb," the *Herald Tribune* added. "Dr. Howell has indicated that certain clouds can best be seeded from the higher level." The company said its planes could get safely airborne even when the police fleet was grounded, and would be ready to fly the city's missions in any conditions except severe icing and thunderstorms.

Smaller than the B-17s and B-29s flown by Project Cirrus, the dual-engine DC-3 was one of aviation's greatest triumphs. Designed and built by Douglas Aircraft, it was a pioneering American passenger airliner during the 1930s. In its heyday, it carried more passengers farther and quicker than any competitor. The later military version, designated the C-47 Skytrain by US forces and the Dakota by the British, flew in every Allied theater of operations during the war. Where the civilian version was gleaming and lovely, its military counterpart looked rather drab and pudgy. GIs nicknamed the latter the Gooney Bird, after big island birds they saw in remote areas of the Pacific.

Paratroopers had leaped from C-47s over Normandy on D-day. The planes also towed gliders filled with troops and equipment. Army pilots flew vital supplies in them from India over "the Hump" into China. Howell undoubtedly had seen scores of them in Italy. A shot-up C-47 once flew with one wing from a DC-2, entering aviation lore as the DC-2½. C-47s later helped break the Russian blockade during the Berlin Airlift. During the Vietnam War, American armed forces would fly a fearsome version of the C-47, a gunship dubbed Spooky or Puff the Magic Dragon. Many DC-3s remain in cargo service around the world even today. If no longer state of the art in 1950, the plane was versatile, capable, and almost indestructible. With Sperry's DC-3s, Howell's air fleet would rank behind only those of the Pentagon and Project Cirrus for undertaking American weather operations.

———————

April began on a Saturday with more rain, which was no April Fools' Day joke. The Passover holiday would begin at sunset for the city's Jews—one million in Brooklyn alone—with Palm Sunday coming the following day for Christians. The 1950 US Census also began that Saturday. (Asked by an enumerator whether he had a bathroom or a shower, the skipper of a barge anchored at the foot of Columbia Street in Brooklyn simply answered, "Bucket brigade.")

Rain continued Saturday afternoon, prompting Commissioner Carney to scrub another planned aerial mission at the last minute. As the *Herald Tribune* explained, "There was just too much water in the wrong places and any man-made rainfall might cause a flood 100 miles northwest of New York." The wrong place was Esopus Creek, which was running high. Howell didn't question the decision, saying his plans were always subject to a city veto. He estimated, however, that he might have prompted an additional 0.1 or 0.2 inches of rainfall. He clearly was growing frustrated at remaining on the ground. "It would have been a worthwhile experiment to see what we could accomplish," he said.

Sunshine lured 225,000 people to Coney Island on Sunday. Light rain coupled with temperatures in the low 60s on Monday generated a generous snowmelt in the Catskills. Esopus Creek continued to run high, further delaying the next rainmaking attempt. Water levels in the Ashokan Reservoir topped 65 percent. In its regular Water Situation tabulation, the *New York Times* reported that on Monday, April 3, the city had 115 days of water remaining at normal consumption levels, or 154 days before pressure failed under the current conservation measures. Even if the Catskills saw normal rainfall through April and

May, reservoirs still were likely to be 35 billion gallons below capacity by June 1.

City employees measured the water levels daily. The city then calculated to 0.1 percent how much of overall capacity was available. An Ashokan Reservoir employee named Virgil C. Gordon had made such measurements for decades. A former New York City policeman, he had started working at the reservoir in 1909. News coverage during the drought now brought him a bit of unexpected fame. "Gordon's job calls for him to take the reading of the elevation at Brown Station, Ashokan Reservoir, every afternoon at 3:30 o'clock," the *Kingston Daily Freeman* reported. "He telephones the reading to the Kingston office of the New York City Department of Water Supply, Gas and Electricity. The Kingston office relays the figure to New York City, where it is added to similar readings from the other reservoirs in the vast system, and in that way the figure showing the percentage of capacity of the whole system is obtained." A newspaper on the Upper Peninsula of Michigan, where readers also knew a thing or two about water, later added, "He telephones his reports: 'Ashokan down eight hundredths of an inch,' or 'up six hundredths of an inch.' . . . 'The newspapers print my reports every day now,' [Gordon] said. 'And I'm enjoying getting some attention for a change. I didn't think my job was important. But more and more I get the feeling that it is.'"

Though the precise readings made by Gordon and his counterparts showed significant water level improvement, Chief Engineer Clark had declared on Sunday, April 2, that "that the need for voluntary conservation of water was as great as ever." Clark added that Howell would remain grounded for "a few days" to let Esopus Creek recede. He also took pains to explain why conditions in the Catskills had prompted the postponement. "This thing just happened to hit us in the Achilles heel of our water system," the publicity-conscious engineer told journalists. The Schoharie Reservoir had only one-seventh the storage capacity of the big Ashokan Reservoir and could fill very quickly, sending the overflow via the Shandaken Tunnel into Esopus Creek, which was running high with its own natural runoff. Area farmers were already opposed to the city's rainmaking scheme, Clark noted. "I don't believe we would have flooded the creek," he said. "But even if it rained without our help, and we tried to make rain, they would say we caused it. Esopus Creek has overflowed many times in the past from natural causes."

Clark's remarks might have been meant as much for Howell, the political novice, as for grumpy farmers. Commissioner Carney, however, gave his meteorologist a public vote of confidence on Tuesday. "I'm taking the wraps off Dr. Howell," he told assembled newsmen. Turning to Howell, Carney then ordered, "Now, you get up there and do your stuff!" What he implied but left unsaid was: *But only when conditions are right and there's no risk of appearing as if we've flooded anybody out.* Howell told reporters he hoped to get airborne again the next day.

The delay at least had given Howell the opportunity to find a suitable spot for his planned Catskill headquarters. It was a caddie house near the first tee of a nine-hole golf course, on a resort atop a mountain at 2,280 feet. The resort, Wolff's Lake Florence Cabins, was about to reopen the course for the first time since 1931. The site was remote, 127 miles by car and 95 by plane from New York in tiny Lakewood, "a locality not shown on ordinary road maps four miles south of Roscoe and about ten miles from the Pennsylvania border."

Unlike the old abandoned resort on Overlook Mountain, the caddie house was in good shape. Howell planned to build a 20- or 25-foot tower for the radar set. "It is planned to have an operations room in the house, and it probably will be staffed with a meteorologist, a radio operator, and a radar technician." Howell told the Associated Press that the site had "an excellent outlook in all directions, particularly toward the Catskills." The *New York Times* marked the spot with a bold X on the weather equivalent of a treasure map. Howell hoped to acquire another site close by, near Tennanah Lake (spelled without the *h* in 1950), for office and lab space. He also now knew, thanks to Weather Bureau negotiations, that in about a week he would receive the long-awaited radar equipment from the US Air Force, on loan to the city for six months. Howell later learned that the radar would be delayed again, to about May 1, and that it would come without a generator to power it. Still, anyone could buy a generator, so the overall situation was beginning to look up.

The meteorologist finally got airborne again on Wednesday, April 5. Howell climbed aboard the Grumman Goose at Floyd Bennett Field wearing a suit and overcoat, hatless, clutching a briefcase. Two Water Department workers and Samuel Schenberg, the supervisor of science for the New York City Board of Education, came along as observers. Their police pilot and copilot took off at 7:01 AM with a 100-pound box of dry ice. The officers wore leather flight jackets; the United Press reported that the plane was "in a race with a towering layer of moisture-laden cumulus clouds passing over the Catskills."

Howell planned for the flight to follow a LaGuardia Airport radio beam to the northwest, get a fix on beams from Allentown and Scranton in Pennsylvania, then turn northward for the Catskill watershed. Nature, however, didn't cooperate. Several cloud layers rose as high as 14,000 feet, completely hiding the ground. The airways also were filled with an unusual amount of static, disrupting radio transmissions so much that the pilots couldn't pick up their beams. Trying to navigate to the watershed by dead reckoning would have accomplished little. Pilots probably would been able to find their way above the cloud deck in one of the promised Sperry DC-3s, but they were still being fitted with dry ice dispensers. For today, at least, Howell's goose was cooked.

At 7:55 AM, "the Civil Aeronautics Administration at LaGuardia Field received word by radio that the plane was turning back because of 'radio trouble,'" the Brooklyn Eagle reported. "It seemed pointless to continue, since we didn't know quite where we were," Howell said. The plane landed back at Floyd Bennett Field at 9:10. Nature, the New York Post commented, was "succeeding better than man in piling up water reserves." With no other flights planned soon, Howell went home to Massachusetts.

Eagle columnist John A. Heffernan wondered what would happen when conditions over the Catskills were perfect for Howell and his police aviators with their bomb load of dry ice. "Shall we have expended some hundred thousand dollars and obtained a few cents worth of water?" he asked. "Or shall we find that artificial precipitation has let loose a torrential deluge, which may tax the dams in the city water system and produce floods where floods are not needed?" With a reference to a figure in Thomas Carlyle's *The French Revolution*, the writer wondered, "Is Commissioner Carney to be New York's cloud compeller?"

Later that day Carney and some of his engineering aides launched a new city conservation program. The commissioner met in his office with Dr. Charles R. Hursh of the US Forest Service. Hursh was a top-notch hydrologist and head of watershed management research at a station in Asheville, North Carolina. The federal scientist, the *Herald Tribune* reported, "explained how to get the most out of rain—man-made or God-given."

Now considered a pioneering forest hydrologist and ecology researcher, Dr. Hursh had studied the flow of water into reservoirs for more than twenty years. During his talks with Carney, he said that a single tree might soak up a thousand gallons of groundwater a year, and that selective cuttings along reservoirs and waterways had saved considerable amounts of water in other parts of the country. He had seen such cuttings along rivers and creeks as much as double the flow of summer water. Conversely, planting other trees and changing soil in certain areas to retain the snowpack during winter months and manipulating the runoff from the snowmelt during the spring also could conserve water.

Hurst's recommended methods, even at their most effective, appeared to offer little hope of quick relief. Even taking chainsaws to a million thirsty, thousand-gallon trees would save only 1 billion gallons of water a year. Any rainy day in the Catskills could easily send several times that amount of water flowing down the green slopes and into the reservoirs. And the city estimated that New Yorkers themselves already had saved 38 billion gallons thanks to the many conservation efforts since mid-October. Still, every little bit helped, and Carney asked Hursh to study the conditions upstate. The expert dove into a study of aerial maps and land conditions in the watersheds. He would deliver his findings to Carney seven weeks later, on May 25, recommending the cutting of some broadleaf trees on public land to let more rain reach the ground. "There are some old hardwood trees in the New York watershed that drink too much," the Associated Press reported.

The hydrologist likewise recommended removal of "worthless and unproductive trees" along waterways and reservoirs, both to conserve water and to improve the general health of the forests. Planting evergreens at high elevations would help conserve the snowpack in the Catskills, but felling them on public lands would reduce disease and insect damage in the Croton watershed. "It would be necessary, according to the report, to organize private land owners to cooperate and to buy some 'strategic areas' now privately owned," the *New York Times* reported.

Hursh concluded that the recommended measures might let 5 to 15 billion additional gallons of water reach the Croton watershed annually, but he didn't offer specific figures for the Catskill reservoirs. Given the already fractious relationship between the city and Catskill residents, the necessary cooperation was very unlikely. Dr. Hursh and his recommendations quickly faded from both the newspapers and civic memory.

11

Mystified City

The Schoharie Reservoir reached capacity and overflowed on April 7, Good Friday. Commissioner Carney, on a three-day tour of the watershed, called his office in Manhattan to postpone any flights by Dr. Howell for several more days. They faced a confusing and complex situation. Overall, the city reservoir system was not quite three-quarters full. Spring had just begun. The full heat of summer—and its annual drain on the water supply—was still months away. "Meanwhile, the city, still short of water in the aggregate, faced a situation in which one reservoir . . . was at 'flood storage' and spilling over, while several small reservoirs in the Croton chain were virtually dry," the *New York Times* reported. The paper blamed the contradiction on what it called *freshets*, an old-fashioned word for spring snowmelt and runoff.

Carney and Edward Clark stood together at the reservoir, watching the precious freshwater cascade down the tall, stair-stepped spillway into Schoharie Creek below. The creek flowed north into the Mohawk River, then southeast into the Hudson. The reservoir had reached its capacity at 10:10 on Thursday night. Between that moment and 8:00 Friday morning, 100 million gallons of freshwater had been irretrievably lost. All of it would eventually flow downstream past New York City, useless to its residents. Carney hoped for cooler weather to slow the snowmelt and halt the overflow.

Chief Engineer Clark told journalists that he and Carney also had visited many of the dozen reservoirs and six lakes that made up the Croton watershed system. Unlike the brimming Schoharie in the Catskills, several of the Croton sources were nearly empty. The Croton system overall was only at 61.5 percent of capacity. To explain the difference from the Catskill levels, city water engineers later speculated that "last year's drought dried out not only the surface, but also subterranean water levels in the region." With the city's fifteenth water holiday scheduled for the coming Thursday, Clark wanted New Yorkers to continue their conservation efforts. "It is because we realize how far we have to go to fill these empty reservoirs that we are calling for another water holiday," he explained.

It wasn't an easy sell with water visibly spilling from the Schoharie. "Nature is providing concrete evidence that the alleged New York city water shortage is nothing but a gigantic hoax as we charged," said John Egan, the Kingston attorney who earlier had sued to stop the city's rainmaking flights. Egan said he would seek a new injunction against what he called "this nonsensical rainmaking experiment" within a few days. "Surely there is no necessity for attempting to make artificial rain when the city cannot capture all the rain provided by nature," he added.

Commissioner Carney got the cooler weather he hoped would slow the spring runoff on Easter Sunday, April 9. Temperatures fell even further than he had expected. "Thousands of Brooklynites braved swirling snow early yesterday to attend a score or more of Easter dawn services marking the most joyous feast in the Christian calendar. . . . Half an inch of snow blanketed Mount Prospect behind the Brooklyn Central Library at Grand Army Plaza, where 500 persons gathered for the dawn service," the *Brooklyn Eagle* reported. It was the first Easter snowfall in thirty-five years, and the reservoirs were now at 75.6 percent of capacity. Carney pointed out, however, that in normal years the city reached that milestone in mid-February, and the snowpack was nearly gone. He cautioned New Yorkers that "the closer we get to June 1, the smaller that 75 per cent grows."

Carney nonetheless prohibited further rain missions until the Schoharie spillway was dry. On Monday afternoon Howell and two police department pilots quietly took the Grumman Goose up over the Catskills for a flight without dry ice. They lifted off at two thirty in clear weather and tested the radio navigation scheme that hadn't worked earlier. Howell considered the flight an orientation, likely his first opportunity to see the reservoirs free of clouds. The Ashokan stretched like a jack-o'-lantern's smile east to west across the top of Ulster County, while the smaller Schoharie wriggled like a fat, glistening snake northward across corners of Greene, Delaware, and Schoharie Counties. The Goose didn't land until a quarter to six that evening.

Howell faced criticism from an unexpected quarter before he managed to get into the air again. It came from Dr. Irving P. Krick of Pasadena, California, who headed the American Institute of Aerological Research and was perhaps the most widely known private rainmaker in the country. A magazine writer later described Krick as the "stormy petrel of the meteorological world." General Electric's researchers would have agreed with the description; they had warned Krick against suggesting that he had any connection with their weather research.

Interviewed in Pasadena but not quoted directly, Krick said that seeding clouds with dry ice from a police plane wouldn't deliver "appreciable results," the *New York Herald Tribune* reported. He added that seeding was effective only through combined air and ground operations; that silver iodide generators on the ground could wring "three to four times greater than normal rainfall" from certain storms; and that New York City should treat its water crisis not as an emergency but as a "chronic state of affairs." Lastly, Krick declared that if the city had begun its rain experiments in December, "definite results would have been achieved already."

Howell shrugged off the complaints, saying there was no scientific basis for Krick's estimate of hugely increased rainfalls. The Water Department had already built silver iodide smoke generators in its own shops, he said, and the rain stimulation project planned to use them. The *Kingston Daily Freeman* later described one of New York's mobile smoke generators: "A truck tows a small trailer on which is carried tanks of hydrogen which is burned to evaporate the solution of silver iodide and release into the atmosphere the minute silver iodide particles which are carried aloft by wind." Howell also had purchased radio equipment valued at $6,500 to link the new ground headquarters with two Water Department station wagons on the ground and the police department and Sperry planes in the air.

As for the project's late start, Howell merely said that "we got going as quickly as possible on a partial basis and the full organization is being brought into line as we go along." He didn't add—although he could have—that the city had only hired him less than two months earlier. Howell ended by saying that the city's methods and findings were public property, and any scientific observer was welcome to study them—including Dr. Krick.

On Thursday, April 13, the same day Krick's criticism reached print, New York's embattled rainmaker took to the sky again, "perhaps to try again what he hasn't been able to do yet—make it rain," the Associated Press reported. Howell said it was only an exploratory flight. "But he took along a little dry ice which he could use to seed cumulus clouds in case he found any. Clear, cold weather was reported in the watershed area." With pilot Patrolman Nicholas Jones and copilot Detective Thomas Williams, he took off in the Goose from LaGuardia Airport at 12:20 PM.

So began Howell's first completed seeding flight over the Catskills. He was unable to say later precisely where they had found and seeded the clouds, only that it was somewhere near the Ashokan Reservoir. Cumulus clouds billowed to about 10,000 feet, a height at which the air was considered thin for rainmaking purposes. "Even so, he said, the clouds were super-cooled, a fundamental requirement, and conditions warranted a try, for practice at least," the *Herald Tribune* reported. Howell later called it a "triggerable cloud formation."

The plane made one pass over the reservoir, landed at Poughkeepsie at about 3:15 to refuel, then went back to the watershed for another two hours. The fliers returned in darkness at 6:42 to the police hangar at Floyd Bennett Field. Howell said later they had seeded the clouds for at least an hour, dropping 100 pounds of dry ice at a rate of about 1 pound per air mile along their track. The seeding run lasted only about twenty-five minutes of the five-hour mission. Snow was flurrying in the watershed area even before Howell had begun seeding, but he was unsure whether his dry ice contributed to anything later.

"It would be completely impossible for anybody to say whether or not we increased the yield of today's snow flurries over the watershed area," Howell said. "On any individual flight, how can anybody tell?" A resident near the reservoir had watched the Goose circle through and around the clouds. "It wasn't snowing until after the plane arrived," said Judge John Lester of West Shokan. "They were flying back and forth into heavy clouds leaving a light white streak as they left. Then it started to snow." The judge added that snow kept falling for ninety minutes after the Goose's departure. State police in Phoenicia, on the other hand, said light snow squalls had swirled all day and that little had changed during the afternoon.

Howell told journalists that he wanted to create a "pattern of records," perhaps ten days' worth of seeding flights to compare with days when there were no flights. "Whatever snow we caused was very light," he said. His first official seeding flight held so little scientific interest for him that he merely listed it later in his final report to the city among every other mission attempted by air or ground in the Catskill watershed. But New Yorkers took more notice when they awoke the next morning to discover what the United Press dubbed a "baby blizzard" covering their streets, parks, and apartment buildings.

"A mystified New York woke up yesterday morning to find unseasonable snow swirling in the air, a blanket of the stuff on rooftops, slush on the sidewalks and the mercury—at 6:45 A.M.—down to 25.2 degrees, a record low for April 14," the *New York Times* reported of the spring-time surprise. The *Brooklyn Eagle* had noted, "The Winter-in-mid-April weather, which caused numerous minor accidents and injuries, came the day after the city's rainmaker had done his first 'seeding' of clouds over the up-State watershed area and many New Yorkers were convinced that he caused today's precipitation."

Snow covered much of the state. Albany had five inches, Schenectady six and a half. At Saranac Lake in the Adirondacks, three hundred miles north of New York City, two teenage boys had gone hiking in the snowfall and gotten lost. They took shelter in a cave overnight, then stumbled out of the forest at two o'clock in the afternoon, cold and hungry but otherwise well. Fifteen miles up the Hudson River from Manhattan, Yonkers got only 1.5 inches of snow, but a trolley motorman there greeted passengers with a cheery "Merry Christmas!"

The snowfall wrecked spring sports schedules. The lacrosse game in Schenectady between Union College and Williams College was canceled. So was the Army-Swarthmore baseball game at West Point. An exhibition game between the Brooklyn Dodgers and New York Yankees was scrubbed, too, along with games in Boston, Philadelphia, and elsewhere. Horace Stoneham, president of the New York Giants baseball club, was personally cross with meteorologist Howell. "Look at that snow, and us with our opening game coming up only four days from now," he said moodily. "That guy had better work out a schedule." A Pittsburgh sports columnist later offered the Giants a suggestion: "Stoneham could set up a battery of anti-aircraft guns on Coogan's Bluff, making it advisable for Dr. Howell to strap on a parachute when he climbs into the cockpit."

The *Suffolk County News* on Long Island ran a puckish piece on its front page. "This is written on Friday morning, April 14th, 1950, Atomic Year Six," it read. "Up on the corner lie piled the bodies of local weather prophets who have been dropping off like flies since early this morning. . . . We expect six or seven feet if the atmosphere doesn't get out of whack from all the flying saucers darting about." The *Long Island Star-Journal* also commented, "The mysterious snow storm in the middle of April was being discussed throughout the state by residents who want to know 'who done it.' . . . Some said Howell must have dropped the ice out of the wrong side of the plane, while others insisted Howell shouldn't be paid his $100 daily salary for producing snow."

An official in little Newtown, Connecticut, later wrote to Mayor O'Dwyer, complaining about the four inches of unexpected snow in his community. "Your damned shenanigans with dry ice did it," A. Fenn Dickinson declared. He thought the city should pay the costs of remounting snowplows on Newtown's trucks. "For the love of mike, Your Honor, will you lay off the dry ice?" Dickinson pleaded. "I had plans for using the unexpended balance of my snow removal account for another purpose, but had not anticipated out-of-state interference."

In New York City, millions of residents forgot about the Easter snow just days earlier and took the unseasonal weather personally. They clogged city hall's telephone lines with complaints about icy roads and bridges, all supposedly caused by Howell's flight. Mayor O'Dwyer himself fielded half a dozen irate calls before eleven o'clock. "New Yorkers were calling it 'Dr. Howell's snow storm' as they slushed about in Spring-time Winter," the United Press reported. "Irate motorists muttered imprecations at 'That Man Howell' as they slithered over icy roads, streets, and bridges; householders complained that it was 'all Howell's fault' as they got out their snow shovels," *Newsweek* added. The unusual meteorological event became widely known as "Howell's Snow."

"There was one man in New York, however, who was beaming with happiness and overflowing with praise for Howell," the *Long Island Star-Journal* pointed out. That man, of course, was Commissioner Carney, who welcomed any form of precipitation. "Carney said Howell deserves congratulations for his 'modest, cautious procedure,' and he said the mayor also was pleased with Howell's work." Howell himself took the event in stride. He obligingly posed for an Associated Press photographer in snow-covered City Hall Park. The meteorologist stood bareheaded, gazing toward the low horizon as if expecting the Goose to materialize any second from out of the clouds. One version of the photo apparently was retouched to make it appear that snow was still falling when the shutter clicked. The image went on the Associated Press news wire and ran in newspapers across the country. One paper in Florida ran it under the headline "DID I DO THIS?"

Dr. Wallace Howell surveys the surprise spring snowfall dubbed "Howell's Snow," City Hall Park, April 14, 1950. *AP Photo / Harry Harris*

It was a legitimate question. *Had* Howell caused snowfall as far away as New York City and Long Island by seeding triggerable clouds over the Catskills? John Aalto, one of the city water engineers who had accompanied him on the headquarters hunt a month earlier, had watched Thursday's mission from the ground. Aalto was "convinced that the airplane seeding 'accelerated' the snowfall," the *Brooklyn Eagle* reported. "He was careful to point out that he offered his opinion as a civil engineer and not as a meteorologist."

Howell, though, believed it was highly unlikely that his seeding had affected the city. He repeatedly mentioned that snow flurries had swirled around the Goose during the mission. "We may have helped to make them stronger and we may not," he said. "It's impossible to tell." Weather Bureau forecasters didn't think Howell had much connection with the urban snowfall, either. "I wouldn't say the cloud-seeding had no effect on the particular snow in New York," a spokesman said in a statement that Friday. "But it is most improbable that it affected the general storm system." The bureau's statement indicated that the storm had "formed off the middle Atlantic coast early yesterday—several hundred miles east of the cloud-seeding tests—and that the storm now centered over the Massachusetts coastal region."

The storm still moved slowly southward. Howell wouldn't let it pass without another attempt at seeding, this time from the ground using the silver iodide smoke generators. Curiously, he wouldn't mention the project's first use of silver iodide in his final report to the city in 1951. Perhaps he considered it only a practice run. Nevertheless, numerous newspapers reported that two Park Department station wagons pulling trailers with smoke generators were in action Friday morning near Cobleskill and East Worcester, north of the Catskill watershed. Aalto directed the operation, keeping in touch by radio with Howell at the Water Department's twenty-fourth-floor offices in the Municipal Building in Manhattan. The two vehicles traveled about fifty miles in dispersing their silver iodide smoke.

As Bernard Vonnegut had shown in his earlier experiments, the effects of silver iodide smoke last longer and potentially extend farther

than seeding with dry ice. Howell said that while it was conceivable
some effects of the new seeding might reach the city, they wouldn't
be noticeable to anyone who wasn't a weatherman. "It might cause a
storm to give rain a few minutes longer, but if it did rain a little longer,
it would clear up all the sooner," he said. Snow had been falling hard
when the experiment began, and Aalto said he thought the smoke had
accelerated the snowfall, but it soon stopped. In the end, no precipitation from the Cobleskill-area seeding reached the city. Nonetheless,
the efforts prompted even more calls to city hall.

Accustomed to standing up for scientists, the *Schenectady Gazette*
opined that New Yorkers ought to be grateful for a little unseasonable
precipitation. "The thoughtlessness of those people illustrates one of the
reasons why some scientists, whose lifework is research, are willing to
give all kinds of advice about the possibilities of rain or snow making
but refuse, so long as they maintain their present jobs to hire out as rain
or snow makers," the *Gazette* grumbled in an editorial. Some papers in
communities less connected to scientific research disagreed. "We think
God caused the snow flurries," declared the *Pittsburgh Post-Gazette*. "We
had quite a lot of them here on Thursday and, so far as we know, not a
cloud was seeded in Western Pennsylvania. Our advice, Doc, is to save
your dry ice and work out a better system of storing and distributing
the rain as God sends it."

Attorney John Egan withheld legal fire from Kingston but also misinterpreted Howell's remarks. "We are sitting tight right now," said the
litigator. "We're not satisfied that Dr. Howell made the snow, although
he seems to take credit for it. As of this moment, it might be natural
snowfall. But if he keeps on working, we may be able to connect him
with it."

12

Jupiter Pluvius

The mixture of suspicion and credulity from Attorney Egan and others regarding Howell's activities wasn't entirely prompted by modern scientific rainmaking alone. Centuries of beliefs, theories, myths, and plain hokum made rainmaking a difficult and sometimes confounding topic. Native American peoples in the arid Southwest, for example, had performed sacred ceremonies for their rain deities long before European settlers arrived. But tribal rain dances now struck many as quaint or humorous, and one condescending wire service piece offered a Hopi rainmaker's advice to the "pale face" who dwelled in "the big village to the east." No similar fun was made of Cardinal Spellman and devout Roman Catholics as they prayed for rain to alleviate the drought.

People had also believed for millennia that heavy rainstorms followed the clang and roar of combat. The Greek philosopher and biographer Plutarch noted the supposed phenomenon in his *Lives* in the second century CE:

> And it is said that extraordinary rains generally dash down after great battles, whether it is that some divine power drenches and hallows the ground with purifying waters from Heaven, or that the blood and putrefying matter send up a moist and

heavy vapour which condenses the air, this being easily moved and readily changed to the highest degree by the slightest cause.

By the American Civil War, weather observers were attributing post-battle rain to the concussive force of massed artillery fire. Edward Powers, a civil engineer from Wisconsin, championed the big-gun theory in his 1871 book *War and the Weather*. "When numerous observers, each independently of the others, arrive at an identical conclusion, in reasoning from facts which they have separately noticed in widely different fields, such conclusion is certainly worthy of respect, and may be assumed to contain the elements of truth," Powers wrote. "Of this nature is the idea under consideration—the belief that rain has been, and can be, brought on by heavy discharges of artillery."

Powers cited dozens of battles in which rain had fallen either during the fighting or soon afterward. These ranged from Ligny and Waterloo during the Napoleonic Wars to Buena Vista and Palo Alto during the Mexican-American War to Antietam and Second Bull Run during the recent conflict that many Americans called the War of the Rebellion. The author acknowledged that rain often *didn't* follow a battle but believed that it was "probably in the manner of the firing, as well as in the amount" that influenced rainfall. Powers attributed battle-induced precipitation mainly to the concussive force of the detonations, which produced "electrical action, and the change or communication of motion." The theory was soon disputed, but it persisted because it sounded logical.

Powers's book fired the imagination of a former federal patents commissioner in Washington, Robert Q. Dyrenforth. A self-styled general who had risen only to major in the Union Army, Dyrenforth successfully lobbied Congress in 1891 to fund experiments to test the concussion theory. In making his case, he cited not only Powers but also US senator Leland Stanford of California. "Senator Stanford, for example, told us of his observations during the building of the Union Pacific Railroad through arid country that rain would often follow blasts in places where rain had never been heard of before, and that since the blasts ceased no more rain has fallen." While relying on such anecdotal evidence, Dyrenforth made a remarkably modern argument for meteorological research. "Now, granting that the atmosphere contains just so much moisture, and no more, suppose we were able, by artificial means, to equalize the distribution—to draw rain to regions which need it, relieving by that amount the surplus moisture in regions where it is not needed!"

Accompanied by Powers and several others, "General" Dyrenforth attempted in August 1891 to make rain using explosives attached to tethered balloons on the C Ranch near Midland, Texas. Newspapers reported mixed results, but Dyrenforth was jubilant. "Not only did we produce rain in abundance but did so under difficulties which were sometimes very discouraging and exasperating," he told a Fort Worth newspaper. A special report from Midland in the same paper told a different story. "Most of the material they brought was intended for mid-air explosions, but because of the winds prevailing here at almost all times it is found impossible to control the kites and balloons containing the explosives, having them go off simultaneously with giant powder upon the ground." The correspondent added, "Before leaving Midland Gen. Dyrenforth donned the traditional cowboy attire, including the high-heel boots. His stay here has improved his health wonderfully."

The odd campaign attracted attention even back East. "It will be remembered that the reservoirs of New York city recently ran low, so that many people who had not used water to any extent in forty years wrote letters to the newspapers complaining of waste on the part of their neighbors," a paper reported in late December.

> At that time Gen. Dyrenforth, who was accused of having committed several thunder showers in Texas, was requested to take his theory and $2,000 worth of dynamite and explode them in the Croton valley. Before Gen. Dyrenforth could figure up the cost of a shower for New York the rain fell gratis upon the just and the unjust, and, while the allowance to the latter was not liberal, we were able to get along on it for a couple of weeks.

Dyrenforth's federal funding was reserved for a test during the dry season. Backed by a Chicago-based investment group, he returned to Texas to resume his experiments in winter 1892 at Fort Sam Houston near San Antonio. Dyrenforth used both balloons and 2-pound shells filled with an explosive called roselite that artillerymen fired from mortars. Alexander Macfarlane, a skeptical University of Texas professor, went out to take a look at this latest battle with the heavens.

"The first thing I have to do with gentlemen of your kind," Dyrenforth told him during a long discussion, "is to convince you that we are not a party of cranks." The effort failed with the professor, who tried to convince one of Dyrenforth's investors that the concussion theory had no merit. "Anyone who has studied physical science," Macfarlane said, "knows that concussion is an oscillating motion of the air which dies away in a very few seconds, leaving as its effect the atmosphere slightly heated." The businessman's reply still resonates today: "We do not care for physical science."

Macfarlane tried again with another of the interested capitalists. The second man replied that he "would not argue the question with me; there was too much money sunk in it." Macfarlane reported that after the twelve-hour bombardment he witnessed, "no satisfactory results were obtained from a practical point of view," but that Dyrenforth and his colleagues "claim that the scientific correctness of their theory was proven beyond a doubt." A headline in an Austin newspaper dubbed the Fort Sam Houston experiments A BOOMING FAILURE.

Still, the experiments generated plenty of publicity, and others wondered if weather modification was possible and potentially lucrative. The following year, no less a figure than financier, investor, and inventor John Jacob Astor IV applied for a federal patent for what he called a "rain inducer," which he hoped would outperform General Dyrenforth's pyrotechnics. The idea was to pump warm, moist air from the ground to a much higher altitude, where it would turn into rain. The patent application included a drawing of a tower built on high ground, with a blower to drive the moist air through a large discharge pipe.

"I never tried the 'rain-inducing' scheme, but it would probably work successfully," Astor said. "The essential principal in my application was different from that which General Dyrenforth endeavored to establish. He tried to induce rain by a series of explosions, and I followed his experiments with much interest. My method is to transfer the air directly from the earth's surface to a high level." The US Patent Office denied the application, because "no mechanical novelty was involved." The office did grant Astor a patent for a new type of bicycle brake.

Tagging along behind the hopeful, the misguided, and the spiritual rainmakers inevitably came the fraudulent. Ranching and farming was always a hit-or-miss enterprise, entirely dependent on the weather. Where rain failed to fall, con men flourished. "A dispatch from Fort Scott, Kansas, says that agents of the Interstate Artificial Rain Company of Goodland, Kansas, arrived in this city last night and to-day entered into a contract with a committee of citizens whereby, for the sum of $1,000, they will produce one-half inch of rain over an area of 500 square miles in three days," an Arizona newspaper reported in August 1892. The agents claimed they had operated in twenty-five locations, failing to produce rain only once. The traveling rainmakers now set up shop in a barn on a high prairie. "No one was allowed under any circumstances to remain in the barn with them." The correspondent ended the dispatch on a hopeful note: "At this hour black clouds are forming and the heavens are illuminated with lightning."

Following the outbreak of World War I, observers again took notice that massed artillery fire seemed to precede the outbreak of heavy rain. Many Americans even believed that the battles in Europe were affecting weather in the United States. The *Washington Herald* tried to set them straight. "The fact that storms have followed closely on bombardments may be accounted for in a very simple way," the paper reported in October 1915. "The territory on which these battles took place has rain about every third day. Now, as a rule important engagements are not started in the rain.

They are not usually begun until a day after the rain. Why then, should it be unusual that rain should follow within a day's time?"

But during so tumultuous a time, it wasn't easy to know what to believe. Shortly before the start of the war, scientist Sir Oliver Lodge had declared that it might be possible to regulate weather by running electrical current through a giant copper ring built around the globe. The idea sounded ludicrous, but Lodge wasn't easily dismissed. One of Great Britain's most distinguished scientists, knighted in 1907 for his scientific contributions, Lodge had experimented with electricity, lightning, wireless telegraphy, the nature of light, and more. Irving Langmuir would have recognized him as a kindred soul, if only for Lodge's invention of a device to disperse smoke and fog. Lodge was a curious individual who, like author Sir Arthur Conan Doyle and other eminent figures of the era, was deeply interested in spiritualism. Not surprisingly, Lodge "found his confreres in science the most determined critics of his convictions." They hardly knew what to make of the weather theory he presented in 1914.

"By placing a copper rod around the earth, parallel the equator, which would discharge millions of amperes, the present climatic vagaries due to the waywardness of the magnetic poles might be corrected," Lodge told a group of electrical engineers in London that January. He also proposed more modest alternatives, such as constructing shorter copper rings around the North and South Poles or discharging electricity into the skies via kites or from mountaintops. "If we want rain, we should send up negative electricity; if we want fine weather we should send up positive electricity," Lodge said.

> The difficulty is to get it high enough up and in sufficient quantity. It must be an expensive experiment, but on the top of a mountain, with kites and balloons, and with proper engineering arrangements for permanently doing whatever is desired in the direction of getting up a sufficiently high tension and discharging an adequate quantity—adopting whatever means is best adapted to that end—something is bound to happen.

Scientists and professors came away from his presentation bemused. "That's too deep for me," a Chicago weather forecaster said of the equator idea, "but if Sir Oliver Lodge said so I can not question it." The outbreak of war in Europe that summer ended any notion of conducting initial experiments. Sir Oliver continued to speak and write after the war and in 1932 received the Faraday Medal from the Institution of Electrical Engineers. Ten days after Lodge died at age eighty-nine in 1940, his shade reportedly appeared during a séance at a spiritual church on Columbus Avenue in New York City. During a somewhat rambling visitation, what the *New York Times* called "the ghost of Sir Oliver Lodge" commented on many things, but whether he still believed in his copper-ring-at-the-equator theory wasn't among them.

One of Sir Oliver's less-respected yet still well-known contemporaries was rainmaker Charles Hatfield, a slightly built, well-dressed former sewing machine salesman from Kansas. Born into a Quaker family, Hatfield operated on the West Coast and in Alaska during the early decades of the century. "He was self-assured and well-spoken, without any of the bluster of a confidence man, a swindler," writes a modern historian. He wasn't a braggart, either. "I don't claim full credit for the downpour," he once said about a particular rainmaking effort, "but I do say that I was responsible for holding the storm in Southern California as long as it stayed."

Many small cities and agricultural groups were well satisfied with the precipitation they had paid Hatfield to produce. The rainmaker mixed large vats of chemicals and released the vapors to waft into the upper air, where "the skies, in an attempt to get rid of the odor, kept pouring down heavy-showers for the remainder of the season," the *Los Angeles Herald* reported in 1906. "Since Hatfield has been traveling about the country he has affiliated himself with Jupiter Pluvius until the two are as one and old Jupiter sits back and takes things easy when the able Hatfield puts in an appearance."

But the Yukon boomtown of Dawson City had fired him for failing to deliver contracted rain, and the US Weather Bureau had publicly warned that what rain had fallen near his operations would have fallen anyway. "It is therefore apparent that the rainfall which was supposed to have been caused by the liberation of a few chemicals of infinitesimal power was simply the result of general atmospheric conditions that prevailed over a large area," Chief Willis L. Moore, a predecessor of Francis Reichelderfer, wrote in 1905. "It is hoped that the people of Southern California will not be misled in this matter and give undue importance to experiments that doubtless have no value." But, as the *New York Times* noted in an editorial, "those who are credulous enough to believe in rain-making incantations would not be convinced by any argument. . . . Faith in rain-making only serves to show how misleading is crude experience, especially when results are viewed through a short-focus, narrow-angle lens of self-satisfied ignorance."

Hatfield was still operating nearly a decade later. In December 1915 he struck a bargain with the city of San Diego, site of the Panama-California Exposition. Hatfield agreed to fill the large Morena Reservoir east of the city for $10,000 or receive no fee at all. The reservoir was a major source of San Diego's water; an intricate, thirty-five-mile flume delivered the freshwater from the mountains into the city. Some said that city fathers granted Hatfield a verbal contract simply to get rid of him. It seemed a safe investment, since the reservoir was only one-third full and had never overflowed.

Hatfield built a twenty-foot tower topped by a platform near the dam. He screened the platform with tar paper and lined it with large galvanized iron evaporation pans that he filled with chemicals. "What his chemicals are, he has never revealed," a newspaper columnist later reported. "The tar paper, collecting heat, causes the liquid to evaporate in the daytime, and at night he applies enough heat to produce the same result. Ascending columns of vapor from the tanks have the power, he says, to attract moisture even to the driest spot." Hatfield used as many as twenty-three chemicals; one city official thought his brew was a compound of hydrogen and zinc, but nobody really knew or ever found out.

Rain began falling in early January 1916. This wasn't surprising, as winter is California's rainy season. Hatfield, though, said he was only getting started: "Just hold your horses and I'll show you real rain." The rain kept falling for days on end, flooding parts of the region. A property owner rescued in a rowboat joked, "Let's pay Hatfield $10,000 to quit." The worst happened on January 28. The Lower Otay Dam broke, unleashing a forty-foot-high wall of water down a valley southeast of San Diego. Another dam withstood the pressure of water behind it, only to have an earthwork on one side give way.

Altogether, eleven people died in flooding amid widespread destruction. San Diego wasn't the only area affected. A hundred and more miles to the north, Los Angeles, San Bernardino, and Pomona all were hit hard, with early damage estimates topping $2 million. Roads and railways washed away throughout Southern California. Some San Diegans held Hatfield responsible for the disaster and threatened his life. The rainmaker claimed he had stopped his rainmaking operations the moment that water spilled over the Morena dam—but insisted that the city pay him for the success of his efforts.

The disastrous flood and the rainmaker's demand for payment made news around the country. San Diego's city attorney feared the municipality might have to pay up, although nothing had been put into writing. "The city attorney further avers that by doing so San Diego acknowledges responsibility for the rain and is liable for damages to the tune of thousands of dollars for the destruction of property by the flood. Wherefore the wailing [and] gnashing of teeth in the common council," a newspaper in Tacoma, Washington, explained in an editorial. "Moral: Never tickle Mother Nature. You may throw the old lady into hysterics."

San Diego wriggled out of paying Hatfield's fee by saying it would compensate him for the rain if he assumed liability for flood damages. The rainmaker wisely declined and sued instead. The case dragged on until 1938, when it was dismissed for lack of prosecution. Lawyer Albert Blaustein later mentioned the San Diego case in his 1950 *New York Law Journal* article about the unclear legalities of rainmaking.

Although the rainmaker never collected his hefty fee, Hatfield benefited from the priceless publicity and continued his career. Thanks to
his ceaseless secrecy, no one ever definitely established whether he was
a serious experimenter, a misguided crank, or a clever bamboozler. As
a Seattle newspaper observed in 1921, when he was busy in the Northwest, "Hatfield has been a rainmaker for 20 years, yet according to the
United States weather bureau, making rain can't be done. Whether or
not he can make rain, Hatfield is living proof of a mighty truth. Because
he believes in himself, he is convincing."

Grand visions and theories continued long after Hatfield's exploits on
the West Coast, even as New York City entered the drought in 1949.
A. M. Low, the well-known British scientist, inventor, and science-
fiction author, wrote in August of that year that the science of weather
control was a very promising infant.

Archibald Montgomery Low was "a brilliant eccentric with a pen-
chant for embracing what, at the time, were dismissed as wild and
improbable ideas." He called himself Professor, having once been an
associate honorary assistant professor of physics at Great Britain's Royal
Ordnance College. But Low was no mere academic crank. He had
demonstrated an early form of television, called TeleVista, thirty-five
years earlier. "Dr. Low admitted that the experiment was expensive and
that at present the invention was perhaps not commercially possible,"
the *New York Times* reported in 1914. Low much later became a top
expert in radio guidance systems for rockets.

In his 1949 article for American newspapers, Low cited the research of
Langmuir, Schaefer, and Vonnegut at General Electric. He opened with
a breezy, futuristic account of the US Congress in the year 2050, approv-
ing a weather budget and allocating periods of rainfall and sunshine.
The writer envisioned self-contained, roofed cities that controlled their
own internal temperature and humidity. "Perhaps fortunately for man,
the evolution of weather and climate control is likely to be fairly slow,"

he wrote. "I say 'fortunately,' for it is potentially fraught with dangers which most people do not yet even conceive." These dangers included international conflicts over weather diversions and the consequences of shrinking ice caps and redirected ocean currents. It was also this article in which Low warned that once humans took control of weather and climate, they would have to assume responsibility for events that would previously have been considered acts of God. He added, "Without some overall authority both nationally and internationally, weather and climate might become disastrous."

Of the many memorable weather figures of the nineteenth and twentieth centuries, it wasn't Low, Lodge, or Dyrenforth, nor even Langmuir, Schaefer, or Howell, who made the most lasting impact on American culture. Instead it was Hatfield, whose career provided the inspiration for playwright N. Richard Nash to create *The Rainmaker* and its protagonist, Bill Starbuck. "Nash's rainmaker is not so successful as Hatfield, but there was a folklore quality to Hatfield that Nash wanted in the play," writes a modern theater historian. The 1954 Broadway version starred Darren McGavin as Starbuck and Geraldine Page as Lizzie Curry; the later film version had Burt Lancaster and Katharine Hepburn in these roles. Hatfield even attended the Hollywood premiere in 1956.

"How do you know I'm a liar?" Starbuck demands in the movie. "How do you know I'm a fake? Maybe I can bring rain! Maybe when I was born God whispered a special word in my ear!" Stage productions appeared in London, Sweden, and Germany; *The Rainmaker* today is listed among the top hundred American plays ever written. The film shows regularly on American cable television, fans reveling in Lancaster's swaggering performance as the character Nash once described as "the phony rainmaker, the sky-borne dreamer who was renegade from reality."

Hatfield followed his trade into the 1930s, and he guarded the supposed secrets of his rainmaking process until his death in tiny, arid Pearblossom, California, in 1958. His family honored his request not to make the news public. The Associated Press later noted his passing in a brief dispatch, which read in part, "News of his death came today—three months after the event—in response to an inquiry from San Diego, the scene of his most successful and most tragic rainmaking effort."

The *New York Times* recalled the San Diego flood while looking back on the checkered history of rainmaking in 2003: "Needless to say, efforts like Mr. Hatfield's gave rain-making a bad name. The field was rescued, to some extent, by the legitimate research of people like Dr. Howell."

13

Combined Operations

Dr. Howell returned to Harvard after the surprise spring snowfall and waited for another opportunity to deploy his air and ground forces. He scrubbed two missions due to poor conditions, but everything finally looked perfect on Wednesday, April 19. Commissioner Carney spoke to a Rotary Club meeting in Brooklyn that afternoon. He again emphasized that New York was "one of the most extravagant users of water" but also applauded millions of residents' conservation efforts.

Howell began directing the city's first joint air-ground operation that evening by telephone from Boston. A correspondent for the *New York Times* was nearly poetic in describing it the next morning: "Under starless skies and on the bleak and windy plateau of the almost deserted airport at Stroudsburg, Pa., another attempt at seeding Catskill-bound clouds to produce artificial rain for New York's watershort mountain reservoirs got under way soon after 8:30 o'clock last night."

Stroudsburg lies in a valley below the Pocono Mountains near the Delaware Water Gap, a hundred miles southwest of Kingston. John Aalto and two other Water Department workers had rolled up to the airport about eight o'clock with two silver iodide smoke generators mounted on trailers, pulled by department station wagons they had driven down from the Ashokan Reservoir. The city workers weren't operating within their home state, much less their own watershed. No one was expecting them and the airport was closed, only the manager still working. The crew set up the generators near the office building. Aalto found two New York state troopers on the site, but they were on other business and left soon afterward. With no Pennsylvania officials on hand to object, the water workers fired up their generator. Remarkably, no jurisdictional dispute ever arose over the incursion.

"Millions of silver iodide particles were released into the atmosphere over northern Pennsylvania . . . in the belief that they would drift northward in the vicinity of the upstate city reservoir system," the *New York Herald Tribune* reported. The wind blew the particles toward the Catskills at twenty miles per hour. The clouds above the airport opened up a few minutes after one in the morning, the rain accompanied by thunder, lightning, and high winds. Howell said later that it was "highly improbable" the silver iodide smoke had caused the downpour, because the wind had carried off the smoke too quickly. Aalto's crew kept pumping smoke into the darkness until two o'clock, when the winds shifted from the southwest to the west, away from the watershed but almost directly toward New York City. The crew then stopped seeding—whether on new or standing orders is unclear. At some point during the wee hours, Howell left Boston to hustle down to Brooklyn and the waiting Grumman Goose.

The meteorologist hoped to take off with 150 pounds of finely ground dry ice at 5:00 AM Thursday morning. His air campaign was delayed—first by a rescue mission (the aviation bureau's primary responsibility), then by low clouds and poor visibility at Floyd Bennett Field. Howell finally got airborne at 9:26. The police department pilots headed toward the watershed but soon discovered that the clouds rose above 15,000 feet. This was too high for an aircraft not equipped with onboard oxygen. Ice began coating the plane's wings and static again caused communication problems. The team turned back before the Goose had flown more than thirty miles from the city. "The weather thoroughly fouled up Dr. Wallace E. Howell's carefully planned double-barreled offensive today on rain-bearing clouds over the Catskill watershed," the *Brooklyn Eagle* reported in a page 3 bulletin, "literally raining out one-half of the water-making experiment and blowing the other away."

Worse, the federal Civil Aeronautics Administration (CAA), the forerunner of today's Federal Aviation Administration (FAA), expressed concern about "misunderstandings" between its airways operation division and the Goose. "It appeared that officials were investigating several discrepancies in flight plans and control while in the airways," the *New York Herald Tribune* reported.

This wasn't how Howell had hoped to begin combined operations. He said nothing publicly, but the Harvard meteorologist could have been forgiven if his thoughts drifted back to the war and the many weather problems encountered by the Fifteenth Air Force in Italy. A steep learning curve was perhaps inevitable when attempting a military-style operation. The *Herald Tribune* ran a large illustration to explain the intricacies of the Stroudsburg seeding. It looked less like a battlefield map than a panel from the sports page, complete with statistics and arrows that might have described the critical plays in a football game.

The good news was that the watershed received a soaking overnight—about 0.8 inches at the Schoharie, a little less at the bigger Ashokan. Six billion gallons sloshed into the reservoirs. Unfortunately, the Schoharie had begun to overflow again just ten minutes before Aalto's crew turned on the generators. Esopus Creek was rising as well. Commissioner Carney announced another temporary halt to rainmaking operations in the Catskills; Howell said he might reconsider his decision not to seed the Croton watershed, where reservoir levels were much lower. Overall, the city's reservoir system was now at 78.5 percent of capacity.

The big question was whether the rain stimulation project actually had nudged the weather. Ever the cautious scientist, Howell wasn't yet prepared to take credit. In Schenectady, Vince Schaefer couldn't say whether the city's campaign had contributed to the rainfall, but he believed that silver iodide seeding was effective when properly done. "I wish we could prove it," he added.

The US Weather Bureau wasn't inclined to say that Howell or Aalto had played much role in the overnight storms. "It was nature's rain," Ernest J. Christie, chief of the bureau's station on Battery Place, said to the *New York Times*. "They may have eked out a few extra hundreds of an inch of rain—let's hope so," he added to the *Herald Tribune*. But many New Yorkers remained skeptical. "Whether Howell intended to increase the downpour or prolong it was not clear. One thing is certain," the *Long Island Star-Journal* commented, "New Yorkers will never know whether their $100-a-day rain maker actually earned his pay because there is no way of telling whether the rain is natural or artificial."

The Schoharie Reservoir was still overflowing on April 24, but Chief Engineer Clark said that the city had almost no hope of filling the interconnected reservoirs by June. "I fervently hope I am wrong, but based on past performances the chances are not good," he said. The Catskill reservoirs were at 92.1 percent of capacity, but the Croton reservoirs stood at just 64.4 percent, placing the city's overall water levels at slightly above 80 percent. A year earlier, a month before the start of the city's horrible, forty-one-day rainless spell, the reservoirs had been only 0.5 percent short of full.

The next day, the Bendix Aviation Corporation's Friez Instrument Division in Baltimore offered to loan New York City meteorological equipment worth $10,000 to $15,000. The scientific gear included "instruments for readings and recordings of weather and cloud conditions, and storm intensity and duration, and a shelter for the delicate instruments." Some was standard, the rest would be custom-built, and the city could use all of it without charge for a year. Howell was grateful, and remarked that some of the equipment was beyond his $50,000 budget.

In further encouraging news, Howell's forces sprang back into action later that same day, Tuesday, April 25. The Schoharie Reservoir stopped overflowing at 3:10 in the afternoon, the city having lost more than

2 billion gallons of water over the spillway since the previous Thursday. Commissioner Carney gave his go-ahead to resume silver iodide seeding. At 8:40 that evening, John Aalto and his Water Department generator crew from Kingston went into action at the Stroudsburg airport. Again, the target was the Catskills.

Howell coordinated the operation from New York City, keeping a close watch on wind and weather reports. A light rain began about ten o'clock, with a steady rain following about an hour later. Aalto's generator spewed smoke until seven fifteen the next morning. The Schoharie Reservoir received just 0.12 inches and the Ashokan nearly 0.31 inches of rain, while the Croton watershed on the other side of the Hudson got nearly 0.5 inches. The Water Department estimated that the city's reservoirs captured about 3 billion gallons of rainwater, more than making up for the loss over the spillway.

The man from Harvard again refused to take any credit but sounded optimistic. "From a quick, casual check, it looks as if watershed rainfall was heavier than in most surrounding areas," Howell said. He thought shifting winds might have carried silver iodide particles over the Croton watershed, which would explain the higher totals there. "There was rain over most of New York State and Rainmaker Howell cautioned that more experiments would have to be made and the results studied before it could be definitely stated that scientific rainmaking does or doesn't make rain," the *Brooklyn Eagle* added. Howell returned to Boston as New Yorkers readied for their seventeenth water holiday on Thursday.

During a lull before the next operation, Commissioner Carney watched as the Waldorf-Astoria Hotel began installing a new air-conditioning recirculation system. The hotel expected to save an astonishing 851,000 gallons of water every day—an indication that the O'Dwyer administration had as much conservation work to do in Manhattan and the other four boroughs as in the remote mountains.

Up in the watershed, meanwhile, John Aalto tried to convince residents that the city knew what it was doing and wouldn't accidentally unleash devastating floods. "Speaking at a regular meeting of the Kingston Kiwanis Club at the Governor Clinton Hotel, Aalto explained that a flood is financially as damaging to the city as it could be to watershed residents," the *Kingston Daily Freeman* reported. Swirling floodwaters picked up impurities that were costly for the city to filter out, Aalto explained. The loyal Water Department worker also defended Howell by declaring, "If anyone can control the clouds, he can."

Saturday, April 29, was cold and damp in Manhattan. Most of the spectators expected at New York City's third annual Loyalty Day Parade on Fifth Avenue instead sensibly stayed home. But thousands marched to proclaim their devotion to the United States and its way of life, including Boy Scouts, fife-and-drum corps, mounted policemen, veterans organizations, and groups of émigrés from countries now under Soviet control. Performer Ethel Merman rolled past in a red convertible, Queen of the Loyalty Parade. A group of horsemen dressed as Cossacks trotted by, representing the Russian Anti-Communist Center. "Later, a large group of telephone operators came down the avenue," the *New York Times* reported. "Some bore signs reading: 'The Kremlin has the wrong number, let's break the connection.'" Mayor O'Dwyer watched from a reviewing stand as the marchers filed past for three hours.

Howell was at home outside Boston, directing a new rainmaking operation. It was a sustained ground campaign under Aalto, running all weekend and into Monday. His small Water Department team fired up their silver iodide generators at 11:25 Saturday night at Tennanah Lake, about fifty miles west of the Ashokan Reservoir. They began seeding in calm air associated with a warm front moving eastward. "The experiment started . . . with a generator streaming blue plumes of silver iodide smoke into the cloudy mountain air to stimulate raindrop formation," the Associated Press reported.

Rain began falling at about eight fifteen Sunday morning, the last day of April. When the wind shifted at around nine thirty, Howell directed the seeding team to move forty miles north to Meredith, New York. "Dr. Howell kept telephone contact with Mr. Aalto at three-hour intervals throughout the night," the *New York Times* reported. "Mr. Aalto, with headquarters at Kingston, received reports from the cloud-seeding unit fifteen minutes after each call from Dr. Howell." Rain and sleet fell over the watershed for three and a half of the sixteen hours that Aalto's ground crew seeded the clouds. "The consensus was that the seeding may have prolonged a natural rainfall," the *Times* added.

The Water Department ground crew headed south of the Catskill watershed on Monday, May 1. The men set up their silver iodide generator at noon at Port Jervis, sixty miles southwest of Kingston. The generator broke down at 2:40 PM, but the crew got it back into operation a little over an hour later. Mobile seeding continued as the city workers drove their station wagon along narrow roads, hauling the generator behind on its trailer. The *Times* reported they went north toward Monticello, while the *Herald Tribune* said northeast toward Middletown. The crew stopped seeding at 4:25 that afternoon. Unlike over the weekend, the area saw little rainfall that anyone could associate with the seeding operation. Rather, rain had fallen only until eleven that morning, after which skies were overcast.

Overall, however, the three-day offensive saw significant amounts of freshwater flowing toward the reservoirs, although much of it would take time to get there. Estimates ranged from nearly 2 billion gallons to more than twice that. Again, Howell took no credit. Being out of the city, he wasn't even quoted directly by New York's newspapers. In his final report to the city in 1951, he would express doubt that much of the silver iodide smoke produced during this offensive had reached the areas where it would have done the most good. The city's reservoirs were at 82.9 percent of capacity on Monday morning. They had been full one year earlier.

14

Marksman's Nightmare

Dr. Howell couldn't yet know whether he was actually stimulating rainfall, but his project was generating showers of publicity—which, like downpours in the Catskills, weren't always welcome. For instance, while the recent rains were gratifying to city authorities, they were less popular with sports fans. City hall fielded a number of telephoned complaints, most of them good natured, from baseball fans who wanted to know whether Howell was responsible for a spate of rainouts for New York's three major league clubs. Spokesmen for the Yankees, Dodgers, and Giants dutifully said they didn't think the city's rain stimulation project had anything to do with the postponements.

Herald Tribune sports editor Bob Cooke wasn't so sure. He wrote a column imagining Branch Rickey, the Dodgers' president and general manager, trying to hire Howell to produce even more rain. "You see, my pitchers are all bothered by sore arms," Rickey explains. "We have an important series with the Cardinals coming up and if it could be rained out, we might win the pennant." When Rickey tells the meteorologist to name his price, Howell says he'll take two Brooklyn stars, first baseman Gil Hodges and catcher Roy Campanella. Aghast, the Brooklyn leader asks why. "That's easy, Mr. Rickey," Howell replies. "I'm a Giants fan." (In fact, Howell didn't care for big-time sports, not even Boston's Braves or Red Sox.)

The grumbling didn't end with baseball fans. Brothers Jack and Irving Rosenthal, owners of an amusement park in New Jersey that lay across the Hudson River from 125th Street in Manhattan, offered to double Howell's fee to $200 per day if only he would stop making rain. "While you have been reticent on taking credit for continuous rain and muggy weather all we know is that since we opened daily operations of Palisades amusement park, N.J., April 22, last, we have not had a single day of sunny weather," the brothers said in a telegram. The offer made news as far away as Miami. Howell declined the offer and suggested that the brothers donate the funds to the Mount Washington Observatory for "research on the constitution of clouds." The park instead hired two self-proclaimed "sunmakers," G. A. Sykes and Edward Twardus, allegedly at $500 a day, to deliver fair weather.

"Dr. Howell cannot operate unless he has clouds," Twardus said. "We'll break up any cloud formations before he can get up in his plane to seed them." "By broadcasting sound, light beams, and electro-magnetic waves, we cause various disturbances which break up the clouds and scatter them," Sykes added. The pair also claimed that, if they wished, they could use an equally complicated but unnamed process to attract clouds and produce more rainfall than Howell's team. GE scientists Langmuir and Schaefer no doubt enjoyed reading Sykes's explanation on the front page of the *Schenectady Gazette*.

Despite a gentle nod toward the humorous aspects of the story, a newspaper in New London, Connecticut, found the Palisades dustup a bit alarming. Its editors warned that unless they were careful, Americans might see "a sort of tug of war between rainmakers trying to grab passing clouds in competition with each other, and also experience trouble with persons who are bitterly resentful of the efforts because their special events are rained out." The key question was whether either rain stimulation or prevention really worked. "Clouds can't be produced at will it is reasonably sure; it is a question whether on the other hand they can be driven away."

Coincidentally, in an interview two days later, Vince Schaefer said he thought that the lightning storms that sparked forest fires and the hailstorms that damaged crops probably *could* be stopped by seeding with dry ice. New research into the formation of ice crystals might even help protect skiers from previously unpredictable avalanches, he added. "But, Schaefer thinks research is not moving fast enough," the United Press reported. Sounding a bit like Irving Langmuir, the GE scientist "chiefly blames some still-skeptical and tradition-bound civilian, government scientists."

Howell kept his head down and continued his methodical approach to stimulating rain. A heavy fog rolled in over New York City early on Saturday, May 6, snarling ground, air, and sea traffic. Only ferries ran smoothly, thanks to the radar the vessels were using for the first time. Howell still was waiting to receive his own radar equipment from the US Air Force. Some of it had by this point reached Mitchel Field, the Long Island air force base named for former New York mayor John P. Mitchel, who was killed in aviation training during World War I. But several vital components were missing.

The weather was somewhat better in the mountains than along the coast. Howell ordered both ground generators into action, again directing operations from his home outside Boston. He gave the go-ahead at eight o'clock Saturday morning, and by one that afternoon both teams were in position. One occupied an elevated spot on New York Route 28, two miles south of Andes in Delaware County. The second set up on "a convenient high point" in Sullivan County, near the Grossinger's resort airport outside Liberty. The teams intermittently generated silver iodide smoke until five o'clock, attacking the Catskill watershed from the west and southwest. Sporadic clouds near Liberty produced no rain, but precipitation fell for a half hour near Andes.

By then, the troublesome fog in the city had cleared. New Yorkers enjoyed the warmest day so far in 1950 as thermometers rose nearly to 80 degrees. Chief Engineer Clark said later that although the Catskills reservoirs were now above 95 percent of capacity, the Croton system remained below 70 percent. The latest dry day on Thursday had flopped, with New Yorkers using 3 million more gallons of water compared with the previous week. Voluntary water conservation in the city was more urgent than ever, Clark said, "since there was virtually no chance that the Croton reservoirs would filled by June 1 and no more than a 'fifty-fifty' chance that the Catskill chain would be full." The next day, Sunday, was "that rare gem of a day Brooklynites have been waiting for between teeth-gnashing and wailing for the last few weeks," the *Brooklyn Eagle* reported. The day was fair and warm, with highs in the low 70s, "and with not even enough clouds for the rain-maker to aim at."

Upstate residents fretted that more soggy weather could disrupt the national holiday at month's end—and they knew whom they would blame for it. Among the groups planning Memorial Day celebrations at the fairgrounds in Rhinebeck, "a suggestion has been made that the chairmen compose a stern and convincing letter to Dr. Wallace E. Howell, New York city's consulting meteorologist, suggesting that he keep his silver iodide out of the skies in the vicinity of Rhinebeck on Memorial Day so that rain insurance and raincoats will not be necessary."

Howell continued doing what he could to explain exactly how the city's rain stimulation project worked. He spoke to Phi Lambda Upsilon, a national chemical honor society, at Brooklyn Polytechnic Institute on the evening of Thursday, May 11. While not intended for the general public, the talk attracted a reporter from the *Brooklyn Eagle*, who watched Howell cheerfully fill a blackboard with charts and diagrams. The journalist made fun of the scientific jargon and his own ignorance of it but covered the basic concepts and was generally positive about the meteorologist. "Dr. Howell makes no pretension to knowing all the answers," he wrote. "He inspires confidence because he willingly admits the uncertainty of his activities." Someone in the audience asked about Catskill residents who either had threatened or already had taken legal steps to stop the rain operations. "If they or their lawyers can prove that I caused the rain," Howell replied amiably, "I'll welcome them, with open arms."

The meteorologist had reason to be upbeat. Earlier that day, Justice Ferdinand Pecora had handed down his decision denying the temporary injunction sought by the owner of the Nevele Country Club in Wawarsing. Though Howell still had no way of knowing exactly how he and New York City were faring in their battles with the clouds and air masses gliding above the Catskills, in the courts, represented by expensive lawyers and with no legal precedents to guide anyone, they hadn't yet lost a skirmish.

Howell's ground crews, like his small air force, sprang into action only when conditions were right and clouds were overhead to seed. One of Howell's generators was back in action on the afternoon of May 15, after the first short-term drop in water storage levels in two months. The team began sending smoke into the sky at 12:50 PM from a point near Ellenville, southwest of the Ashokan Reservoir. A light rain began falling locally twenty-five minutes later; at 2:00, sprinkles dampened the streets in Kingston. At 3:00, Howell ordered his second team to begin generating smoke from a high point across the Hudson in Fahnestock State Park. "The mobile generations [*sic*], releasing their silver iodide smoke, were in operation until about 10 p.m. yesterday, the crews operating on both sides of the Hudson River," the upstate *Binghamton Press* reported. (The headline was a classic: IODIDE SQUIRTS CLOUDS, CLOUDS RETURN SQUIRT.) By 1:30 AM, 0.03 inches of rain had fallen over the watershed.

By this point New York State was experiencing a damp spring but not a particularly rainy one—in fact, even the Catskill watershed had seen somewhat less rain than during an average year. Still, many residents in New York and adjacent states remained unhappy with the "cold, misty, and foggy weather that has dampened . . . spirits for about a month." In a joint news conference on May 17, both Dr. Howell and Ernest Christie of the local Weather Bureau office tried to convince people that the city's activities were simply an indicator of the weather pattern, not the cause of it. "I am pretty darn sure that there is no connection between the two," Howell said. "As far as the produce farms in New Jersey and the week enders in Westchester and near-by Connecticut are concerned, there is no connection." Even if rainmaking operations had influenced the weather, he added, "we do know certainly it was restricted to the Catskills and the immediate vicinity." The Associated Press also quoted him as noting that "only God can bring the clouds."

Howell would speak the following week to an evening meeting of the Hudson Valley division of the American Institute of Electrical Engineers at Poughkeepsie. The hosts opened the event to area farmers and local mayors. Everyone had the same question: Was the recent rain due to Mother Nature or to Wallace Howell? "I wish I knew the answer," the meteorologist replied. "It's like a marksman's nightmare, in which he shoots at a target while other men are shooting indiscriminately. You take one shot and find six holes in the target. Then a reporter asks you if you hit it."

Commissioner Clark, meanwhile, imposed another short ban on Howell's cloud-seeding operations on Thursday, May 18. State officials had asked for the pause so that traffic experts could make surveys for the proposed New York State Thruway, which would parallel the west bank of the Hudson River near the Catskill watershed. Without naming names, the *New York Herald Tribune* reported that some "rain-making enthusiasts . . . viewed the department's orders for a cessation of seeding during the traffic count as over-cautious." The state workers finished just in time. Cold rains sent an estimated 3 billion gallons of runoff streaming toward the reservoirs on Friday, May 19. With a second traffic pause planned for Sunday, Howell swung quickly into action when conditions looked favorable for seeding on Saturday. He flew down from Boston at the last minute to oversee combined air and ground operations.

Howell took off in the Goose from LaGuardia with a police department pilot and copilot at 1:15 that afternoon, carrying 150 pounds of dry ice. He planned to drop the payload along a line between

Middleburgh and Coxsackie, north and east of the Schoharie Reservoir, respectively. At about the same time, one of his ground generators began working somewhere in the watershed (the newspapers later neglected to say where). The Goose returned at 4:45 with 100 pounds of dry ice still on board. The meteorologist told journalists he had seen two showers east of the watershed when the Goose arrived, but none over the reservoirs until he had begun seeding from between 8,500 and 9,500 feet. He conceded that the watershed precipitation "appeared to be the result of seeding" but backtracked as scientific caution kicked in. "I don't know if the seeding caused it," Howell said. "I doubt we made a great deal of rain at most."

Three days later, on Tuesday, May 23, the meteorologist launched another combined operation in and over the Catskills. For the first time, he flew the mission in one of the Sperry DC-3s. "The plane has greater range and altitude capabilities than the Police Department's Grumman Goose which has been used on previous seeding forays," the *Herald Tribune* reminded readers. "It is outfitted with experimental equipment for simplified navigation in bad weather and for blind landings, permitting operations in weather when police planes might be grounded."

The Goose delivered dry ice to the Sperry airfield on Long Island that morning. The DC-3 then flew north to meet Howell at Poughkeepsie, where he had spoken to the electrical engineers the previous evening. The crew took off for the watershed at 1:30 PM. The target area was Ellenville, southwest of the Ashokan Reservoir. A mobile ground crew already had begun generating silver iodide smoke at 12:40, moving from Wurtsboro to Port Jervis on a direct southwesterly line with Ellenville, away from the reservoir. The team continued making smoke until 4:00, then moved northwest to Narrowsburg on the Delaware River, along New York's border with Pennsylvania, where they finished work at 7:00 PM.

Howell stayed aloft in the DC-3 until 6:00, when the plane landed safely at LaGuardia Airport. The crew had scattered the dry ice from 12,000 feet—much higher than Howell ever had managed in the smaller, unpressurized Goose. "Asked if he had caused any rain,

Dr. Howell said he thought some rain had been produced," the *New York Times* reported. Again, the methodical scientist wasn't about to claim any great victory over the heavens—accurate reporting took time and careful calculation. The Harvard scientist might have said even less had he known that this would be his only flight on a Sperry DC-3.

15

Cloud Pirates

Howell's planes weren't the only ones seeding clouds over the Catskills in the spring of 1950. Freelance rainmakers had begun taking to the skies in the months and years following Vince Schaefer's flight over Mount Greylock in November 1946. By the turn of the decade, squadrons of private planes were dropping granulated dry ice over New York and much of the United States, flown by pilots dreaming of rain, a quick payday, or both.

Things had quickly grown flinty in the American West, where water was precious and cloud seeding instilled both hope and fear. Nick Gregovitch ran cattle in the Huachuca Mountains of Arizona, down near the Mexican border on a ranch he called Nicksville. He had read about Schaefer's groundbreaking experiment in the December 30, 1946, issue of *Life* magazine. Small springs on his land were low and the grass roots were dry. With his livelihood dependent on weather, he had nothing to lose in trying to duplicate Schaefer's Massachusetts snowfall. One afternoon four days after reading the article, Gregovitch took off in a light plane he had acquired in lieu of collecting a debt, carrying dry ice originally used to pack ice cream. He climbed to 10,000 feet and began dumping his surplus dry ice onto the clouds below.

"Five minutes later he was struggling to keep his little plane right side up in a swirling snowstorm," *Life* itself later reported. "Late that evening when he drove back from the distant field where he had had to land, he found his ranch glistening under a 5-inch blanket of snow. 'Hell,' said Nick, 'this is a good deal!'"

The rancher kept seeding for months, neighbors helping to pay for his dry ice and fuel. Gregovitch painted NICK'S in white letters atop one of his sheds, to guide any other fliers who got lost in storms over his place. According to *Life*, the grass range was thriving by August. The magazine dubbed Gregovitch "Nick the Rain Maker." The *Arizona Republic* newspaper in Phoenix noted in an editorial that *Life* hadn't mentioned the many skeptics who doubted the effectiveness of dry ice rainmaking. Both publications posed a key point: If rain was no longer an act of God, what laws or moral principles applied to clouds whose moisture may have been bound for other areas? "Nick Gregovitch, his neighbors who have shared the expense and benefits of summer rains over their parched range, and others who have watched clouds race past to areas where rain is needed less than in Arizona, may have the answer to that nebulous question," the *Republic*'s editors wrote.

The quotable Arizonan in the cowboy hat popped back into the news that winter because of another rancher, Richard Haman, whose spread lay south of Reno, Nevada. Haman filed a claim in December 1947 with the Nevada State Engineer's Office in Carson City. He firmly believed that the water in all the clouds passing over his property near the Nevada-California state line rightfully belonged to him. "He was not cloud-cuckoo-land-crazy," *Time* magazine reported. "He intends, he explained, to sprinkle dry ice on some of the clouds, and he wants full title to the rain he may bring down, wherever it falls." As Haman himself explained, "We plan to make rain . . . and we want to make sure we have full legal rights to the water we produce."

Gregovitch was having none of it. Now leader of an Arizona cloud-seeding group, he immediately kicked up a fuss. He charged that Californians and Nevadans were trying to steal Arizona's share of the water that flowed into the Colorado River. "California wants to hog the water of a river into which it doesn't put a drop," Gregovitch said, "and now one of its yes men wants to hog the clouds." The Arizonan pledged that if Haman somehow won legal rights to the clouds that drifted over his property, "I'll drop loops over them before they get there."

Efforts by the Reno Chamber of Commerce to make snow for skiers at Mount Rose likewise brought complaints. Neighboring Utah objected to Reno's attempts to "milk our clouds" during a season when snowfall was less than usual. The *New York Times* warned of "an undercurrent of seriousness in the wordy 'battle of the dry ice' which the residents of Utah and Nevada are waging with the aid of press communiqués." Back east, at a meeting of farm groups a month later, Vince Schaefer worried aloud about uncontrolled cloud seeding with dry ice. "One can easily visualize considerable trouble ahead, if this type of 'cloud pirating' were uncontrolled," he warned.

Groups and individuals of all types and abilities had studied General Electric's 1946 experiments and decided to try rainmaking themselves. "As a result there was a burgeoning of new cloud-seeding efforts initiated by commercial operators, industrial organizations, water districts, and groups of farmers," a US Senate committee reported decades later. "Some used ground generators for dispensing silver iodide obviating the need for airplanes and their attendant high costs, so that many such operations became quite profitable. Many rainmakers were incompetent and some were unscrupulous, but their activities flourished for a while, as the experiments of Schaefer and Langmuir were poorly imitated."

While Nick the Rainmaker was busy in southern Arizona, the owner of the Topeka Owls minor league baseball club in Kansas threatened to sue the local *Topeka State Journal* over the newspaper's plans to seed rainclouds with dry ice. His lawyer wrote that the owner "has a right to the free and uninterrupted use of the hot and dry atmospheric conditions which he has learned to love and enjoy here in Topeka without being molested by busybody rainmakers with high falutin' methods." The *State Journal* promised to seed only during the daytime so as not to interfere with night games.

In Colorado two weeks later, an amateur flier was pleased with the brief showers he claimed to have produced by seeding clouds with 4 pounds of dry ice—until he got home to his wife, irate over her now-soaked washing that had been drying in their yard. But rainmaking reports didn't always include claims of such pinpoint accuracy. "T'other day Chickasha, Okla., residents sent up a dry icer, the wind came up, and it rained gloriously in nearby Anadarko," the Associated Press noted. "So there you are. Even when you do something about the weather, you still can't depend on it."

A crop duster in Kansas City also had read about Vince Schaefer's experiments and begun his own rainmaking activities on a whim. He professed to be the first flier to offer set fees for seeding with dry ice. "Actual flying time is at the rate of $45 an hour from the time his plane leaves the home airport," the AP reported in December 1947. "That is a base rate. In addition his schedule calls for $100 for a shower that will settle the dust; $500 for a soaking rain from ½ to one inch, and $1,000 for a run off rain of one inch or better."

Rainmaking wasn't strictly an American craze. In New South Wales, Australia, people blamed the wet summer (which coincided with the Northern Hemisphere's winter) on a suburban Sydney man's experiments with cloud seeding. "Some of the correspondents charge 'the rain-maker' with having perverted the laws of God, and say that the community is being made to suffer for it," the Sydney Morning Herald reported. Others suggested that the activities were a "form of modern witchcraft." The rainmaker defended what he called "only simple scientific experiments." He added that they were similar to tests recently conducted by Australia's Council of Scientific and Industrial Research.

And then came the private rainmaking flights over the Catskill watershed that coincided with Dr. Howell's official efforts. Alfred Rudich, operator of the airport in Monticello, claimed to have made three seeding flights over Sullivan County. The first was on March 21, 1950, a week before Dr. Howell first got airborne in the Goose. The others were April 1 and April 5, before Howell had yet completed any seeding missions. Rudich's aerial weapon wasn't dry ice but silver iodide, which Howell would deploy only with ground generators. Rudich and a former navy pilot flew a Vultee BT-13 Valiant, originally a single-engine, two-seat military trainer. Their surplus Valiant "shoots silver iodide after it has been combusted with a spark from the plane's engine," the Associated Press reported. The Monticello man claimed that his flights had produced two snowstorms and a heavy rain. Rudich's approach to rainmaking was unscientific but far ahead of most scientists in dispersing silver iodide particles from the air.

Many amateur and professional rainmakers were using ground generators as well, especially in the West. The units were simple to construct, as Commissioner Carney's workers had proved by building a pair in the Water Department's shops in New York City. So many rainmakers began using silver iodide that by the spring of 1950 scientists feared that seeding was beginning to interfere with legitimate projects, such as the Project Cirrus experiments then underway in New Mexico. The arid state enlisted the New Mexico School of Mines to research ways both to produce rain during a continuing drought and to prevent unsanctioned attempts at rainmaking from making conditions even drier. "The action came after Dr. Irving Langmuir of Schenectady declared belief that wildcat rain-making attempts are making the drought in the southwest worse," the AP reported in late May. *Time* magazine later noted Dr. Irving Langmuir's denunciation of commercial rainmakers, "many of them woefully ignorant of the art, who are seeding the atmosphere with silver iodide throughout the dry Southwest. 'Some of them,' he said, 'are using hundreds of thousands of times too much. No more than one milligram of silver iodide should be used for every cubic mile of air.'"

Langmuir's General Electric colleague Bernard Vonnegut would openly speculate in August 1950 about licensing rainmakers. "Many farmers, ranchers, and civic minded people in many parts of the country are now engaged in cloud seeding," he told a meeting of the New England Association of Chemistry Teachers at Storrs, Connecticut. "In their efforts to produce more rain, these amateurs are releasing large quantities of seeding material which may well contaminate the atmosphere so as to hopelessly confuse the more careful experimenter and precipitation analyst."

Such contamination would affect not only GE's work but New York City's rain stimulation project as well. The *Schenectady Gazette* would loyally agree with Vonnegut, declaring in an editorial that rainmakers had "sprouted like mushrooms in various parts of the country. . . . In the absence of good arguments to the contrary, we would suggest area control—that is, by agreement among groups of states. New York and New England, for example, might enter a compact for identical control of rainmakers' activities."

Despite the experts' concerns, rainmaking continued to grow in popularity, soon becoming so commonplace that, according to federal government estimates, from 1948 to 1952 commercial operations grew to cover 10 percent of the continental United States. Seeding with dry ice and silver iodide continued throughout the 1950s and well beyond. The subject even popped up years later in a 1965 episode of *The Dick Van Dyke Show*, in which young Richie Petrie dresses as a cloud for a science class. The boy explains it all to neighbor mom Millie Helper:

> RICHIE: I'm a good cloud. An airplane flies over me and
>   seeds me.
> MILLIE: Seeds you? What in the world for?
> RICHIE: To make me rain on the crops.
> MILLIE: Listen, my Freddy's gonna be an ear of corn. You
>   gonna rain on him?
> RICHIE: Sure. You want him to grow, don't ya?

No one asked the suburban New Rochelle boy whether his good cloud also would help provide fresh drinking water for nearby New York City, which that same year would be facing yet another drought.

16

Weather Headaches

On May 26, 1950, the Friday before the Memorial Day weekend, Dr. Howell sent one of the Water Department's two silver iodide generators into action. He was relying more on his generators as the weather warmed. With John Aalto again supervising the operation, the crew set up on high ground near Millerton at the edge of Dutchess County. This was a rare instance of Howell seeding clouds over the Croton watershed, across the Hudson River from the Catskills.

At the time, Memorial Day was still celebrated not on the last Monday in May but on May 30, which that year fell on a Tuesday. New Yorkers thus had a four-day holiday weekend to look forward to, and people all across the region were clamoring for Commissioner Carney to help make it a success by suspending rainmaking efforts altogether. Carney said that such a move would serve no purpose, since the meteorologist went to work only when conditions were already right for rainfall. "Dr. Howell's activities cannot spoil the weather," Carney again insisted. "He can operate only to increase precipitation when there are already clouds for seeding."

On Saturday, the city's rain stimulation project got unwelcome criticism from a worldwide authority on trees. During the 1930s, British forester, author, and activist Richard St. Barbe Baker had encouraged his friend President Franklin D. Roosevelt to create the Civilian

Conservation Corps. He was also noted for his work in Africa. St. Barbe Baker was now on his way to receive an honorary degree from a school of forestry in North Carolina. "I feel that it's a sacrilege to bombard the clouds," he said. "It's not fair to your neighbors. You give water to those who don't want it." Howell and Carney didn't reply, but upstate newspapers were happy to print the Briton's remarks.

Howell ended up ordering no operations over the long weekend, though small amounts of rain still fell on the Catskill and Croton watersheds. By Monday morning, the Ashokan Reservoir was 99 percent filled. "The east basin can still hold an additional one billion gallons before overflowing, the New York Department of Water Supply said," the *Daily Freeman* reported in nearby Kingston. "The west basin has been overflowing into the east basin for some time." New York's overall water system was at just above 90 percent capacity, with the Croton reservoirs lagging behind those in the Catskills.

Though the city's rainmakers were idle over the holiday, rain did end up interfering with some New Yorkers' weekend plans, washing out a Monday game between the Giants and the Dodgers. Baseball fans hoped for more cooperative weather on Memorial Day itself. "With the permission of Dr. Wallace E. Howell, the energetic rain-maker, and the co-operation of the elements themselves, all three local baseball teams today will resume action after yesterday's washout by coming to grips with one of their deadliest foes in the major league pennant races," the *New York Herald Tribune* reported on Tuesday. The weather held, each club played a doubleheader, and they collectively came away with five wins. The Yankees and Dodgers swept the Red Sox and Phillies, respectively, at home, while the Giants split with the Braves in Boston. (The paper's horseracing writer would likewise nod to the meteorologist the following weekend. "Thanks to Dr. Wallace E. Howell, rainmaker-in-chief for New York City who must have taken the day off, the skies were sunny and the afternoon was pleasant yesterday at Belmont," he wrote. "It should also be added that some of the prices were healthier than the weather.")

The ground generators were back in action on Wednesday, the last day of May, amid spotty showers over the Catskill watershed. The mobile crew again moved from Wurtsboro to Port Jervis, then continued four miles farther south. Silver iodide smoke drifted skyward from noon to eight that evening, Aalto reporting moderate showers at Port Jervis. Down in the city, water officials were upbeat about the water in the Catskill and Croton systems, which were now collectively at 91.2 percent of capacity, "far better off than had been thought possible at the low point last December . . . of 33.4 percent, representing near-disaster proportions," according to the *New York Times*.

Commissioner Carney gave considerable credit to water-thrifty New Yorkers. Water Department figures put their conservation savings at 55 billion gallons, without which the reservoirs would have been under 70 percent of capacity. Statisticians calculated that the water saved was "sufficient to create a lake 201 feet deep over all of Central Park, or put all twenty-two square miles of Manhattan under twelve feet of water." The water rise, Chief Engineer Clark said, was "a remarkable achievement."

But the *Brooklyn Eagle* warned that a dry June would put New York back in the same position as during 1949, when water levels had continued falling throughout the second half of the year. "Still," the paper added, "New Yorkers had pulled themselves up by the bootstraps in five months, with the aid of Mother Nature and some questionable assistance by the city's rainmaker."

The next day, June 1, Commissioner Carney announced that several of the city's bans—on watering lawns and gardens and washing the outsides of buildings, and on swimming pools using recirculating water—all would be eased at midmonth for thirty-day trials. Parks, botanical gardens, housing projects, and other big public areas would be able to sprinkle two days a week, but that was all. "Anyone thinking of using a hose on a hard surface this summer can drop that idea right now," the commissioner warned. As June began, the surface of the New Croton Reservoir was still seven feet below overflowing.

"Oddly enough, the month of May, despite what seemed like a constant drizzle, failed to produce its normal quota of rainfall," the *Eagle* noted in an editorial. In contrast to its occasionally snarky reporting earlier, the paper added, "And efforts at rain-making, we think, should not be dropped, particularly if conditions late in the Summer and next Fall indicate a return of empty reservoirs."

———————

June 1950 proved an important but ultimately disappointing month for Howell and the rain stimulation project. His ground crews seeded clouds with silver iodide smoke six times between the first and the twenty-ninth, operating from New York and eastern Pennsylvania. The only official combined air-ground operation came on June 3, when the meteorologist and two police pilots took the Grumman Goose and seeded clouds with 20 pounds of dry ice from 15,000 feet, on a line between the Ashokan Reservoir and Turnwood, New York, about thirty miles to the west. The flight was turbulent, the Goose at times hitting winds of sixty-five miles per hour. The Harvard meteorologist "popped an aspirin in his mouth when he landed at LaGuardia Field," the *Eagle* reported. "Like many a Brooklynite, he had a weather headache." Howell couldn't say later whether the flight had produced any rain. "We're not coming down through clouds over mountains to see," he said.

The big Ashokan Reservoir finally reached capacity the same day, sending 10,000 gallons a minute over its spillway and on toward the Hudson. A feature column titled It's a Strange World in the *Binghamton Press* ran comparison photos of the now-filled reservoir and "an almost empty supply . . . about six months ago." Howell knew how such articles must strike readers and again tried to explain why it was important to continue the rain project. "Even if the reservoirs are overflowing," he said, "we want to find out if rainmaking will be valuable during low water periods."

Howell made what the *Herald Tribune* called a "personal cloud-seeding flight out of Albany to the Catskill watershed area" on June 17, apparently not in the police department's Goose or a Sperry DC-3. The clouds he sought had already moved on by the time he arrived. He didn't know it at the time, but he wouldn't again take to the air with dry ice for New York City. The meteorologist later expressed general satisfaction with the month's results. He thought the pattern of rainfall matched up well with the direction of the smoke plumes from his pair of silver iodide generators; he specifically would mention four of the June storms in his final report to the city in 1951.

The watersheds were now receiving enough runoff that Cardinal Spellman decided New York's Catholics could stop praying for rain. *New Yorker* magazine had earlier published a cartoon showing two clergymen watching raindrops flow down an ornate church window. "I wonder if it's theirs or ours," one says. Cardinal Spellman had his answer. "His Eminence said it was evident that God in His goodness and mercy had deigned to answer the prayers of His humble servants by granting sufficient and wholesome rain," the archdiocese announced on June 11. The cardinal expressed his gratitude for his parishioners' response, "which he felt had helped in no small way to avert a threatened and disastrous water shortage." The cessation perhaps wasn't soon enough for fifteen hundred Water Department employees who sailed up the Hudson River on June 15 for a summer outing at Bear Mountain. Heavy rain kept them on the boat; Commissioner Carney "denied, however, that Howell was responsible."

Upstaters' prayers, or at least their wishes, were still for the rain to go away. Farmers and agricultural agents bitterly complained about too much rainfall and called for a stop to New York's activities. Commissioner Carney politely turned aside all such requests. "Despite superficial appearances, the recent rainstorms in the East have developed and progressed in an entirely normal fashion, without any widespread features even remotely attributable to the city's experiments," the Water Department said in letters responding to the appeals. "Your attention is invited to statements to this effect by the Weather Bureau." Carney provided some relief to New York City residents, however, by making the Thursday dry days monthly rather than weekly events.

The angry upstaters didn't surrender. Sixteen officials from the Monticello area, the "saddest of the hate-Howell groups," descended on Manhattan on June 22 to see what they could do about lessening the rain. They arrived on a DC-3 on the first of the summer's daily short-hop flights between the city and the Catskill resorts. Upon landing at LaGuardia Airport, Monticello mayor Luis de Hoyos promptly sounded off about Wallace Howell. "He's raising hell up there," he said. "We know that rain can be made by the seeding of clouds. There are always clouds in the mountains. That man is making it miserable for us. We want relief."

The delegation met with New York City Council president Vincent R. Impellitteri over lunch. Asked how the tourist season was progressing in the Catskills, de Hoyos shot back, "What season? We're all rained out so far." Impellitteri "squirmed but bravely carried on as Mayor O'Dwyer's greeter of distinguished visitors," the *New York Times* reported in an article headlined CATSKILL OFFICIALS JUST LOATHE HOWELL. "The City of New York is thankful for the cooperation of the officials of Monticello in this rainmaking experiment," Impellitteri said, as diplomatic as a foreign minister. Then he broke into a grin. "Don't forget," he cracked, "you're getting all this rain free of charge." Howell would probably remember the joke later, during the winter, when Impellitteri wielded far greater power and influence over the city's rainmaking.

By the third week of June, New York's water situation continued to improve. The Water Department calculated that consumption was now 961 million gallons a day, down from 1.334 billion gallons during the same week in 1949. "The difference between those two figures is more than one-third of a billion gallons," Commissioner Carney said, jubilant over the dramatic drop. "If the public will stay with us and hold down consumption this way, we can avoid the trouble we got into last fall."

At the end of the month, as summer officially began, Carney eased the restrictions on wading pools and street and playground showers. "We are only permitting these operations so as to give children every chance to enjoy the pools and showers during hot weather," he said. But kids across the five boroughs couldn't yet frolic in the fanned spray of open fire hydrants, and continued loosening of water restrictions depended upon the unreliable cooperation of Jupiter Pluvius.

Howell couldn't feel as cheerful about the rain project as city officials. He had always envisioned a radar-equipped headquarters high in the Catskills, from which he would direct the airborne dry ice campaigns, and he had spent his early days as a consultant hunting for the perfect site. As gray June passed, events began to dispel the plan, slowly, like valley mist under unexpected sunshine. Poet T. S. Eliot was wrong: for the Harvard meteorologist, the cruelest month wasn't April but June.

The first problem still was securing an appropriate site. The owner of the Lakewood tourist cabins had withdrawn his offer to lease the city his unused caddy house. Howell renewed his search. On June 5, he chose a hilltop between Callicoon and Livingston Manor in Sullivan County, east of the Delaware River. The property totaled eighty-eight acres, but Howell wanted only one-quarter of an acre on which to set up his equipment, laboratory, and radar tower. "It has been offered to the city for $80 a month under a lease calling for six-month occupancy with an option for the city to continue using it for an additional eighteen months," the *New York Times* reported. Howell relied on the real estate bureau of the city's Board of Estimate to negotiate the lease.

The next challenge for the headquarters plan was getting the long-awaited US Air Force radar equipment. American forces had used advanced APQ-13 radar during the war in the Pacific theater on B-29 bombers. Many sets were converted to weather use after Japan's surrender. The gear intended for Howell still waited, incomplete, at Mitchel Field. Howell had no idea that in the end it wouldn't matter, because the third and final problem was insurmountable.

The news appeared in every newspaper in the country on Sunday, June 25. WAR IS DECLARED BY NORTH KOREANS; FIGHTING ON BORDER, read a front-page headline in the *New York Times*. "SEOUL, Korea, Sunday, June 25—The Russian-sponsored North Korean Communists invaded the American-supported Republic of South Korea today and their radio followed it up by broadcasting a declaration of war."

The peacetime armies of the United States and South Korea were almost completely unprepared for the invasion. They soon give ground. Enemy forces pushed them to what became known as the Pusan Perimeter, where bloody fighting would rage into the fall, until General Douglas MacArthur's daring end-around landing at Inchon on September 15 eased the pressure. The Korean War would seesaw back and forth before ending in a stalemate in 1953.

Howell, the World War II veteran, certainly understood what this new war in Asia meant for his rain stimulation project. With American and United Nations troops and matériel surging toward Korea and staging areas in Japan, New York City never managed to obtain the missing parts for the APQ-13 radar set, and air force technicians were no longer available to operate it. (Project Cirrus also lost most of its military aircraft and personnel.) Howell even lost one of the journalists who had covered his search for a Catskills headquarters site. Bob Eunson left for the Associated Press office in Tokyo, there to be the bureau chief.

Worse, the big Sperry DC-3s were no longer available for cloud seeding, as the planes and crews were siphoned off to Korea. Howell was back to relying on the NYPD's little air wing.

17

Summertime

Whether the conditions weren't right for seeding or city officials had quietly urged him to briefly suspend operations, Dr. Howell didn't order either of his generator teams into action during the July 4 holiday, another four-day weekend. With the arrival of summer, New York again was consuming more than 1 billion gallons of freshwater every day for the first time since December. The weather was generally humid and warm but not scorching, and the occasional showers were not enough to make up the daily water deficit. Storage levels in the reservoirs were slowly dropping.

After the cool, dismal June, New Yorkers made the most of the opportunity to enjoy the outdoors. On July 3, seventy-five thousand people passed through the gates of Palisades Amusement Park in New Jersey—the largest Monday crowd since 1940 at the park whose owners had recently tried to pay Howell an exorbitant fee not to make rain. Across the nation, Americans took to the roads and highways in staggering numbers. Nearly five hundred died in car wrecks, a record total that a safety official called "sheer slaughter." The wonder in New York was that an equal number weren't accidentally crushed on the area's many beaches. On the Fourth of July, several million people swarmed to the seaside, toting blankets, beach balls, and plastic buckets. A million and a half went to Coney Island alone; during the day sixty-three lost children were returned to frantic parents and fourteen swimmers were rescued by lifeguards.

Weather across the region abruptly changed that afternoon. An erratic northwest wind brought rain and hail to Westchester and Putnam Counties. A seventy-year-old Queens woman died when her car skidded and crashed during the storm. At 3:40 PM, the tempest began toppling trees on Long Island. Six racing sailboats and other small craft capsized on the sound. The sailors perhaps should have known to put into shore, since many motorists had heeded radio and newspaper warnings to start for home early, "many leaving their weekend retreats at noon and early afternoon." Others extended their holiday to head home early Wednesday, "creating from 8 to 11 o'clock this morning one of the heaviest volumes the parkways have handled on the day after a holiday," a Yonkers newspaper reported the next day.

The generator crews likewise returned to work on July 5, again directed by Howell from Boston. They would send silver iodide smoke drifting skyward seven times during July, their busiest month so far. Howell again was generally satisfied with the results. His teams still operated only when meteorological conditions were right for stimulating rainfall over the watersheds. Jupiter Pluvius seemed now to be reversing New York's frightening drought a year earlier. "Rains that approached a tropical intensity yesterday drenched the metropolitan area, bringing benefits to some and trouble to others," the *New York Times* reported on July 11. "The storm was general throughout the Middle Atlantic region."

The sporadic storms for once drifted more often over the Croton reservoirs than those in the Catskills. "The heaviest July rainfall in the watershed area in the last 50 years today provided a huge gain in reservoir supply for the fourth straight day," the *Long Island Star-Journal* happily reported at midmonth. Nearly 6 inches of rain had fallen during the month on the Croton watershed—three times more than in 1949—raising water levels in the reservoirs there to nearly 85 percent of capacity. Rainfall over New York City's watersheds was 8 percent higher than average since January 1. Storage in the reservoirs rose during the first two weeks of July for the first time since 1935.

Howell directed his ground crews to keep seeding, just as he had been hired to do. The Water Department gave no indication that it wanted him to stop or slow down. To the contrary, the city observed the first monthly dry day on Thursday, July 13. (Chief Engineer Edward Clark did allow Bond Clothing Store to reactivate the waterfall in its signage on Times Square—the water was recirculated, so none was wasted.) "Even though we have announced continued relaxation of water restrictions for another thirty days, we ask the people to go on with their conservation efforts so that when the summer is over it will not be necessary for us to reimpose these restrictions," Commissioner Carney said. Howell continued to direct his generator crews remotely, staying out of the limelight and rarely seeing himself quoted in the metro newspapers. His biggest fan was his son Stephen, back at home outside Boston. Whenever he saw rain, the six-year-old ran to his father and asked the same question: "Did you do it, Dad, did you?"

Others remained far less delighted by the Harvard rainmaker. On July 10 the Sullivan County Board of Supervisors unanimously passed a resolution calling his rainmaking operations a "public nuisance" that the big city should stop to "avoid future ill feeling and legal activity." The resolution warned that the rainmakers "may be subjected to grand jury inquiry and possible indictment and prosecution." One irate member even said the board should have passed not a resolution but an ultimatum. "New York should be forced to stop the practice," he said. "Sullivan County is suffering and should not be hampered as a resort area."

The Orange County Board of Supervisors passed a similar measure four days later, also unanimously, demanding that the city "desist" from more rainmaking operations in the Catskills. This second resolution declared that the city's cloud seeding "might very well be responsible" for the heavy rains that had damaged vegetable crops on truck farms in the county, a farming area situated between Sullivan County and the Hudson River. Later in the month, the Liberty Chamber of Commerce passed yet another anti-rainmaking resolution. The editor of the *Liberty*

Register wrote to the editor of the *New York Times* explaining that the "rains which have fallen upon the Catskills and nearby Schoharie, Dutchess, and Orange Counties throughout the spring and much of the summer have hit hard at the economy of the resort counties and have retarded the normal growth and harvest of crops in the strictly farming counties."

The three resolutions echoed the festering public resentment in the Catskills. A newspaper near Livingston Manor had reported threats to take potshots at Howell's aircraft with deer rifles. The Associated Press later reported an ominous remark by a farmer in Dutchess County, where people blamed rain from seeded clouds for washing insecticide spray off their apple trees: "I know some people up there that would shoot Dr. Howell on sight." Fortunately, there were no known instances of anyone firing shots. A farm bureau agent in the same county, however, believed that Albany should step in. "State control is needed for rainmaking," he declared. "It's not fair for the city to make rain during the hay-making season."

A common complaint was the threat to tourism in the region. "The Catskills, the greatest vacation area in the world, is adversely affected by the New York rainmaking propaganda. Incalculable harm has been done," declared a columnist in Margaretville's *Catskill Mountain News*. "The question of whether or not Dr. Howell, of the rainmaking forces, can really make rain is not the point. The gist of the matter is the advertising." To the writer, the issue was one of perception: with cloud seeding getting widespread publicity throughout the region, tourists couldn't be blamed for hesitating to visit area resorts. "The result has been that thousands of people seeking vacations in the Catskills have been scared away by the rain publicity and the entire mountain region has suffered severe losses," the paper added in a later article. "Mountain businessmen believe that New York could have at least held off on the rainmaking for July and August—the vacation season."

Several small Catskill newspapers took the same tack as upstate lawmakers and demanded an outright halt to the rain stimulation project. The *Kingston Daily Freeman* even attached a comment to a normal Associated Press report on the Weather Bureau's long-range forecast. The paper ran the piece on the front page beneath a head-line that asked, in part, WHY DON'T YOU LET US ALONE, RAINMAN? It compounded the slight by misspelling Howell's name—probably unintentionally—in an unusual preface. "Attention Dr. Howells, New York Rainmaker: The following accreditation on the weather is offered by *The Freeman* as a public service. What you intend to do to upset the forecast is not known to us. How about taking a long-delayed vacation during this period?"

Catskill residents were correct about the barrage of publicity devoted to the city's rainmaking project. It was a modern story that captured readers' attention. The metro newspapers kept tabs on the shifting water levels in the watershed and reported every operation by Howell's air or ground teams. Though the meteorologist was now loath to offer quotes to the press, the mere mention of his name was "as portentous and pervading as the storm clouds that have dark-ened New York's watershed for most of this summer," according to the *New York Times*. Readers had been conditioned to accept the notion that Howell's activities were indeed inducing rainfall over the Catskills—probably well beyond any actual effects they had—by reams of articles about the rainmaking phenomenon, published in local newspapers and popular weekly and monthly national magazines ever since Vince Schaefer's flight over Mount Greylock in 1946. The articles often were optimistic and uncritical, reinforcing popular per-ceptions that amazing, never-before-imagined realities were possible through modern science.

Meteorologist Lawrence Drake and other skeptics believed that many articles misinformed the public. "Since the A-bomb, our journalism, nimbly leaping the chasm between hopeful hypothesis and provable result, has been transforming 'science' into a new magic, complete with myths and miracle workers," Drake wrote in 1948. Like Schaefer, Drake had a diverse and colorful background, but his critique hadn't been appreciated in the hometown of Schaefer, Vonnegut, Langmuir, and the House of Magic. The *Schenectady Gazette* had noted his criticism beneath the headline "RAINMAKERS ARE ALL WET"—So SAYS EGG FARMER.

Langmuir's free publicity machine churned out fresh copy during that damp summer of 1950. In the July 14 issue of *Science*, the Nobel laureate described various methods of seeding cumulus clouds. The technical article, complete with mathematical formulas, dealt mostly with the results of silver iodide seeding in New Mexico. Langmuir concluded by stating his belief that seagoing silver iodide generators one day might be used to stop hurricanes from churning ashore.

All of that was mindboggling enough, but an assertion about dry ice had newspaper editors across the country reaching for their blue pencils. "The use of a single pellet of dry ice," Langmuir wrote, "about a 3/4-inch or 1-inch cube, may have distinct advantages for seeding. . . . Often, heavy rain may best be obtained from a large cumulus cloud by using a single pellet of dry ice shot into the side of a cloud from a Very pistol." The *New York Times* ran a short piece about the *Science* article under the headline EASY TO MAKE RAIN, LANGMUIR ASSERTS. It was little wonder, then, that angry Catskill residents blamed Howell and his air and ground forces for bringing down ruinous weather on their resorts.

The city's rain stimulation project did have defenders other than Commissioner Carney. A letter to the *Times*, signed Pro Bono Publico, disputed the earlier letter from the *Liberty Register* editor by citing years when far more July rain had fallen than in 1950. "In those years was the resort business ruined?" the writer asked. "No. Were all crops destroyed by too much rain? No. One could not sue nature. But one can sue Dr. Howell and New York City. A suit that was a failure."

Syndicated New York newspaper columnist Billy Rose also defended the meteorologist, humorously praising his unwillingness to make extravagant claims of rainmaking success. Unhappy upstaters misjudged the situation, in Rose's view. "There's a persnickety perverseness in folks which makes them believe the opposite of anything you tell them," he wrote. "If Hopalong Howell had claimed credit for the rise in our reservoirs, chances are the newspapers would have ganged up on him and called him a double-domed phony." And Howell's becoming modesty was so novel among New Yorkers, Rose added, that all eight million residents "are kind of stuck on him."

It was high summer in the city. Water consumption rose with the heat, and perhaps due to assumptions that there was less need to conserve after the recent rains. Storage levels in the reservoirs had dipped for eight straight days when Howell's teams began seeding rainstorms with silver iodide smoke on August 1. One generator worked for eight and a half hours at Grossinger's airport outside Liberty, the other for six hours between Margaretville and Turnwood. Commissioner Carney reiterated that "water use in the generally hot, dry month of August probably would decide whether water curbs must be reimposed by the end of summer."

With his six-month contract with the city soon due to expire, Howell spoke with a journalist at the Municipal Building on August 7. The ever-cautious meteorologist finally made his first claim of success. "Dr. Wallace E. Howell, the city's consultant meteorologist, came right out yesterday and said it: 'I have made rain,'" the *New York Times* reported. Howell then dampened the moment by adding that he had produced "very little rain that would not have fallen anyway." He even acknowledged a "statistical possibility" that in some instances his activities had prevented or reduced rainfall, although he hadn't seen any evidence of it. When a *Herald Tribune* reporter reached him by phone, Howell "showed no signs of excitement. . . . He said that on occasion he had been able to watch clouds from which he had made rain but insisted that 'on those occasions the results were very slight.'" Despite the careful phrasing, most readers and journalists remembered only four words: *I have made rain.*

Thursday, August 10, was the city's second monthly dry day. Storage levels in the reservoirs had fallen for twenty-seven consecutive days and

stood at 87.1 percent of capacity. Commissioner Carney used the occasion to announce a thirty-day suspension of city rainmaking activities.
The pause, he said, would give the police department time to upgrade
the cloud-seeding planes. The *Herald Tribune* reported that oxygen
equipment was to be added; the *Times* reported that it was silver iodide
equipment. Carney added that Howell would use the time to complete
a report on his first six months' activities. This report was "likely to
satisfy few persons," the *Schenectady Gazette* noted in an editorial. "His
observations leave a wide open field for both pro- and anti-rainmakers.
Protesters against rainmaking can point out that he admits he made
rain. Defendants can point out that it's not certain he made rain that
wouldn't have fallen anyway."

Carney may well have initiated the monthlong suspension to
temporarily relieve the pressure from those demanding that the city
curtail Howell's rain stimulation work. But the commissioner's early
skepticism was long past, and he had already said he wanted Howell
to continue his work. "The Commissioner now requests," the city's
budget director wrote, "in order that the rain-making experiment cover
the four seasons of the year, and in order to obtain maximum benefits
from such experiments, that Dr. Howell's contract for directing the
artificial stimulation of rain and snow from the clouds be continued
for another six months."

In other words, Carney still backed the Harvard scientist, despite
continued yelps from upstate. This was all the support that Howell
realistically could hope for, and probably more than he would have
received from any other major city. The Board of Estimate granted
Howell the requested contract extension on August 17. It didn't need
to appropriate additional funds, since the project had spent less than
half of its initial $50,000.

"Certainly Dr. Howell has made only the most self-effacing com-
ments," the *Herald Tribune* stated in an editorial. "But the man's reputation is nevertheless made. We hired Dr. Howell, we got the extra rain,
and what more do you want? Perhaps Dr. Howell should be allowed
to write his own ticket at this point."

18

Señor O'Dwyer

William O'Dwyer was no longer New York City's mayor. On Tuesday, August 15, President Harry S. Truman had named His Honor the United States' ambassador to Mexico. The first hint of the appointment had hit the newspapers only twenty-four hours earlier. "During the past month the Mayor, it was learned, has been quietly taking care of some of his 'loyal stalwarts' with quiet appointments in well-paying patronage jobs as a prelude to his quitting the City Hall post," the *Brooklyn Eagle* reported. The same edition contained another article, under the headline COURT TURNS BACK RACKET JURY REPORT. In New York City, with its complex and rough-and-tumble politics, the two items naturally were connected.

A month earlier, what the Associated Press later called "the biggest police-bookmaker graft scandal of recent history" had broken open, with headlines perfectly suited for tabloid journalism. The sordid tale involved a reported $1 million in annual graft, paid to police officers for protecting a bookie's $20 million gambling ring. Captain John G. Flynn, who commanded one of the divisions under investigation, had testified in June before a grand jury. The following month he shot himself in the head with his service revolver and died in his own station house. Kings County (Brooklyn) district attorney Miles F. McDonald launched an inquiry, and Mayor O'Dwyer, who'd been both a police officer and a gang-busting D.A., sided with the men in blue and publicly clashed with McDonald.

The *Eagle* thought the mayor had become "obsessed" with the investigation over the summer. "Finally he came out into the open with an attack upon it as a 'witch hunt,' seizing upon the suicide of Police Captain Flynn as evidence of the alleged mistreatment of the police by District Attorney McDonald," the newspaper stated in a long editorial. A grand jury rejected the mayor's claims, finding that the investigation was "fair, fearless, and honest." A charge that Flynn had been the subject of "cruel psychological torment," jurors added, was "hysterical and without basis." Eighteen officers would later stand trial in the case, which collapsed when the bookie, headed for jail anyhow, refused to testify against them. Although no charge was ever filed against O'Dwyer, the investigation permanently stained his mayoral legacy.

Unhappy, in poor health, and under mounting political pressures, O'Dwyer got out while he could still salvage some of his reputation. "It is hardly surprising that a man whose spirits rise or fall like a thermometer should finally become irked with the daily grind of a tough routine in the mayor's chair. . . . We hope he enjoys himself south of the border," the upstate *Utica Daily Press* editorialized, under a headline that referred to him as "Señor O'Dwyer." "And, boy—are some Democrats glad to see him go where he can watch bull fights—instead of being a bull in the party china shop." The mayor and his wife slipped quietly away from Manhattan on the evening of Monday, August 14. They rode south on the same train as New York Yankees outfielder Joe DiMaggio, although they didn't run into him until reaching the station in Washington.

President Truman was no fan of the mayor, who had worked against him during the 1948 Democratic convention. But politics were politics, and Edward J. Flynn, a Democratic National Committee member from the Bronx (no relation to the late Captain Flynn), hammered out a deal. The mayor arrived at the White House at twelve fifteen on Tuesday to spend just twenty minutes with the president in the Oval Office. O'Dwyer didn't stay for lunch but later called at the State Department.

"Yet there was a tinge of resentment apparent among diplomatic career men that Truman was replacing one of their number with what they consider a political appointee," the Associated Press reported. "O'Dwyer told newsmen at the State Department late in the afternoon he regards the appointment as a great honor." Truman left it to his press secretary to announce the country's new ambassador to Mexico.

Back in New York City, news of O'Dwyer's imminent departure sparked a flurry of speculation and maneuvering. The only modern mayor to resign there had been colorful James J. "Jimmy" Walker, under pressure from then governor Franklin Roosevelt in 1932. Columnist Robert Ruark wasn't alone in finding the current situation highly peculiar. "All of a sudden our beamish broth of a bye [*sic*] breezes off to Mexico. Did he jump or was he pushed? Was it heart trouble, cop trouble, gambling trouble or the firm fist of Democratic Boss Ed Flynn, or a combination of all?" Ruark asked. "Whatever it was, Bill's subjects don't like the smell."

Election laws and the calendar now steered events. The mayor said he would step down on the last day of August. Because he was resigning before September 20, voters would select a new mayor in a general election on November 7. City Council president Vincent Impellitteri—the bureaucrat who had met and tried to soothe angry upstate mayors during the summer—would serve as acting mayor through December 31. If O'Dwyer had stayed beyond the critical date of September 20, Impellitteri would have been mayor at least one additional year, through an election for a two-year term in November 1951. Instead, a small army of Democratic county committeemen would now nominate a mayoral candidate, their choice affecting Democratic Party races across the state.

"For example, should Mr. Impellitteri be nominated for Mayor—on the basis of an early resignation—this would rule out Ferdinand Pecora, also of Italian ancestry, for consideration for Governor, it was held," the *New York Times* reported. The Pecora reference was to the same New York Supreme Court justice who had ruled in the city's favor in the rainmaking lawsuit. The *Times* also noted later, "If, as now seems likely, [party leaders] decide that a Protestant is to be the nominee for

Governor, and Herbert H. Lehman of the Jewish faith accepts as scheduled renomination for United States Senator, the mayoralty nomination will probably go to an Irish-American of the Roman Catholic faith." The *Brooklyn Eagle* noted several other possible hats in the mayoral ring, including the city controller, Manhattan district attorney, and Brooklyn borough president—as well as Commissioner Carney. On the Republican side, speculation swirled around US congressman Jacob K. Javits.

As first in the line of succession and the soon-to-be acting mayor, Impellitteri might have expected his party's nod for the permanent job. Democratic leaders thought otherwise. Kingmaker Flynn declared a "wide open race," the *Times* noting that Pecora "appeared to rate the party endorsement ahead of Mr. Impellitteri." No firm decisions were expected for a least a week. O'Dwyer, meanwhile, bade a teary farewell to the city's Board of Estimate. "He was an immigrant, the city took him in, set him down at its fireside, and gave him all the opportunities it gives to all of its own children," the mayor said of himself. "It gave him honors above his deserts. That same city now is launching him on another assignment."

The following week, on August 23, the *Eagle* reported on a conclave of the city's five Democratic county leaders at the Hotel Commodore. The news was discouraging for Impellitteri. "The 'deal,' reportedly being made between the Tammany organization and Bronx Democratic czar Edward J. Flynn, involves the choice of Supreme Court Justice Ferdinand Pecora as the Mayoralty nominee and Bronx Representative Walter A. Lynch as candidate for Governor," the newspaper reported. Subsequent events showed that the report was right on the button.

———

On August 31, the city said good-bye to Mayor O'Dwyer. Five thousand soldiers, sailors, airmen, marines, and city employees marched from the Battery along lower Broadway, the Canyon of Heroes, which had seen many parades. Oddly, O'Dwyer wasn't a participant. The parade reached city hall at twelve thirty, shortly after the mayor and his wife arrived from the mayoral residence at Gracie Mansion, in the opposite direction. An estimated crowd of fifty thousand people—twenty thousand of them city workers—listened to O'Dwyer's farewell address. He delivered it, the *Brooklyn Eagle* noted, "at a not far distance from where George Washington made his farewell address to the people of the United States."

O'Dwyer first echoed his emotional remarks to the Board of Estimate, then began a litany of his administration's accomplishments. Sounding like a newly minted general at the start of the Civil War, he said he had been summoned by the president of the United States for a vitally important job. "The post to which I have been called," he said, "offers a fine opportunity to serve the cause of democracy in a world threatened by the menace of totalitarian aggression." O'Dwyer concluded with a *beannacht leat*, an Irish blessing: "May you continue to progress toward a better and happier life in accordance with the best traditions of American democracy."

Vincent Impellitteri, not yet officially the acting mayor, presented O'Dwyer with the mayoral flag as a keepsake. The departing politician and his wife then rode away in a motorcade to a farewell luncheon. Police estimates for the total size of the various crowds ranged from one-quarter to one-third of a million people. "It was not, however, an enthusiastic or noisy crowd," the *New York Times* reported. "There was polite applause for the Mayor's speech and moderate handclapping as his open car moved through the streets, en route to the Waldorf-Astoria, stopping occasionally to permit the Mayor to shake hands with old friends, mostly members of the Police Department."

The O'Dwyers departed the city by rail on the 20th Century Limited at six o'clock that evening, bound for a stay in California before heading to Mexico City. The mayor left behind a one-sentence letter for the New York City Board of Elections: "I hereby resign as Mayor of the City of New York." The resignation took effect at ten thirty Saturday morning, September 2. Impellitteri, still the city council president, was now also the acting mayor until after the November election, with the full powers of the office.

The acting mayor wasted no time settling in. "I am moving into Gracie Mansion next Tuesday and I expect it to be my official residence after November 7," Impellitteri said. Though O'Dwyer had made numerous last-minute appointments, his successor refrained from shaking up the administration. "I have no present plans to remove anyone," he said. A week later, on the evening of Saturday, September 9, nine thousand Democratic county committeemen (representing a total of thirty-eight thousand) filled the main floor and two side balconies of Madison Square Garden to select their party's new mayoral candidate. The choices had boiled down to the acting mayor and Justice Pecora, now considered the favorite.

Impellitteri, age fifty, had come with his family to America as a child from Italy. He had served in the US Navy as a young radio operator on a destroyer during World War I. After earning a law degree in 1929, he served as an assistant district attorney, secretary to two state Supreme Court justices (neither of whom was Justice Pecora), and two terms as city council president. The *Times* decades later recalled Impellitteri as "a slightly built, courteous man with little of the turbulent energy or whimsical humor" of O'Dwyer or LaGuardia. "His speech was devoid of colorful language and his gestures seemed limited to clasping his hands or slowly twisting one of the dozen cigars he smoked daily." Historian Robert Caro later described him as "amiable but slow witted . . . a joke among political insiders." He added, "But now he was mayor—and the joke was on the city."

Ferdinand Pecora was nearly a generation older at age sixty-eight. Also born in Italy, he was a widely respected jurist who had spent a dozen years in the district attorney's office before serving as counsel to the US Senate's Banking and Currency Committee under President Roosevelt. He had taken his seat on the bench in 1935. Pecora now had the blessing of his party's power brokers, and he counted Commissioner Carney among his supporters.

A former state Supreme Court justice placed Pecora's name into nomination. The party's choice was "well known in advance, and the only question was whether Acting Mayor Impellitteri's name would also go before the house," the *Times* reported. A little-known Brooklyn committeeman did nominate Impellitteri but mispronounced his name twice in doing it. The acting mayor's backers complained of "steamroller tactics." The meeting began late, at 8:40 PM; Pecora secured the nomination at 10:03. The justice invoked both Roosevelt's New Deal and Harry Truman's Fair Deal in his acceptance speech. "I intend to continue many of the fine services inaugurated in our city since World War II under the last administration," Pecora said.

Mild-mannered Acting Mayor Impellitteri left the Garden determined to knock the Democratic Party on its ear. He had warned that he would mount an independent candidacy if denied the nomination. Backers headed by real estate executive Walter T. Shirley had already formed a citizens' committee, headquartered at the Chatham Hotel. "I certainly don't intend to repudiate the spontaneous support of my candidacy by a substantial group of business people, headed by Mr. Shirley," Impellitteri had declared.

The acting mayor's backers swung into action following the Garden debacle, collecting signatures to put his name on the November ballot as a candidate for the newly created "Experience Party." Their efforts made Democratic Party bosses nervous, especially after Impellitteri spurned offers to nominate him to the state Supreme Court instead. One Impellitteri booster, Mrs. Florita Rogovoy, set up folding tables in Times Square to collect signatures. "You would be surprised how many Republicans have signed my petitions for Impellitteri," she said. She also claimed that Tammany Hall officials had appeared three times, tearing up placards and telling her to "lay off." The feisty gray-haired supporter refused. "I'm going to do this until I get a million signatures," she vowed. Impellitteri needed just seventy-five hundred, but all within a week. He filed over sixty-seven thousand at the deadline on September 16.

The ballot choices for mayor in the November election were now Justice Pecora, Democratic and Liberal Parties (nominated by both); Impellitteri, Experience Party; Edward S. Corsi, Republican Party (another Italian immigrant, Corsi had held several state positions in the administration of Governor Dewey); and Paul Ross, American Labor Party (Ukrainian-born, Ross had been a city official under O'Dwyer). The smart money backed Pecora, Corsi had little chance, and Ross had none. Impellitteri, the wild card, thought he would win.

The acting mayor at least looked the part of a chief executive, and he had excellent public relations men to reinforce that image. Crucially, he also landed the endorsement of Robert Moses, the city's construction coordinator. A former Republican candidate for governor, Moses was among the most powerful and Machiavellian figures in New York City's long history. "Impellitteri privately promised Moses even more of a free hand than he had enjoyed under O'Dwyer in setting all city construction policies," Robert Caro later wrote in his Pulitzer Prize–winning Moses biography. Moses came out for the acting mayor two weeks before the election. "This is a community of good sportsmen," he said. "New Yorkers like a fighter and especially one who starts under a handicap without organized support and refuses to be seduced, bought, or frightened off."

One of Impellitteri's targets during the mud-filled campaign that followed was the Board of Water Supply. Its members, he charged, were "appointed for life under a vicious mandatory law which gives the Mayor no control whatever." Despite two terms as the city council president, Impellitteri positioned himself as the outsider candidate. He was "the Man Who Cannot Be Bought, Bossed or Bribed," his newspaper ads proclaimed, adding, "Impy's Got the Bosses on the Run!" The election promised to be more entertaining than the funny papers for New Yorkers. But with its attacks on the Board of Water Supply, and so by implication Commissioner Carney and other O'Dwyer allies, the Impellitteri campaign was a dark cloud on Wallace Howell's horizon.

19

Autumn

Dr. Howell spent his monthlong imposed break writing his report for the city summarizing the first six months of rainmaking operations. Safely out of action, he couldn't be blamed for a "torrential downpour which flooded Brooklyn under almost three inches of rain" and sent water rising countertop-high in parts of Queens on the weekend of August 19–20. Upstate farmers were now split in their opinions about his rainmaking efforts, a change that Howell might have considered an improvement. "Sometimes they like it and sometimes they don't," the *New York Times* reported from a Farm Bureau Federation meeting at the Dairymen's League offices on Park Avenue. "Some farmers think that Dr. Wallace E. Howell, New York rainmaker, has helped their crops. Others feel he's dug a watery graveyard for their hay."

After the storm in the city, storage levels in the city's watershed reservoirs began to fall again, as they usually did in the summer. They declined for twenty-three straight days and had slipped below 80 percent by the time Howell got back to work. The meteorologist arrived in Brooklyn on the afternoon of September 13, ready to test a newly installed silver iodide generator in the police department's Grumman Goose. "The planes, not on a rain-making run since June, have only used the dry-ice method to date," the *Herald Tribune* reminded readers.

Howell likely wasn't surprised when he again faced the problem that had bedeviled him from the start—too much rain. "About 2 P.M. yesterday Dr. Howell arrived at Bennett Field from Boston to try out the police plane. He waited patiently all afternoon and finally gave up the effort at 5 P.M.," the *New York Times* reported. "Incessant rain and low visibility prevented a take-off." Howell said he would try again "as soon as conditions make it possible."

He finally took to the air in the Goose the next afternoon, on his thirty-sixth birthday, making a "dry run" with the silver iodide gear. "He said that a few minor adjustments would be made to the apparatus and that he hoped to give it a real rain-making test over the watersheds in a few days." Howell later added, according to the *Times*, that "with the approach of colder weather, air operations would be more effective than cloud-seeding from the ground." But it wasn't to be.

While perfect for police operations, the venerable Goose had proven less than ideal for cloud-seeding missions during their previous efforts with dry ice. And Howell apparently disliked the new airborne silver iodide dispenser, though he never explained why. The mechanism was more complicated than the chute that dumped the dry ice; perhaps it simply didn't work well enough to suit him. In addition, having lost the well-equipped Sperry DC-3s, and never having obtained the radar to direct his flights to pinpoint locations, Howell no longer needed a headquarters in the Catskills. His ambitious plan for a mountain site had already vanished from the newspapers. Now his entire air campaign quietly disappeared as well. Howell and his police pilots never flew another seeding mission. From this point on, city teams operated only from the ground, generating silver iodide smoke.

———

September 21 would be the city's twenty-seventh and final dry day "until further notice." More good news came from Commissioner Carney, who announced that most water restrictions were now permanently removed. "Praising the public for saving some 75,000,000,000 gallons of water since last December, Commissioner Carney nevertheless appealed for continued conservation," the *Brooklyn Eagle* reported. The watershed reservoirs were at 79.1 percent of capacity, a huge improvement from 60.2 percent on the same date in 1949. Although pushed from the front pages by news of the daring amphibious US Marine landing at Inchon, South Korea, on September 15, the drought appeared broken at last.

Howell and Carney had no intention of ending the rain stimulation project, however. Its purely scientific component was still important. Howell's generator teams went back into action the same day the Americans stormed ashore at Inchon, operating from elevations west and southwest of the Ashokan Reservoir. They sent their silver iodide smoke into the skies just once more in September and three times in October. Howell was especially pleased with the results of seeding during a storm on October 9–10 from Bear Mountain, near the Hudson River about sixty miles south of the Catskill watershed.

Irving Langmuir also popped back into the news with remarks at the annual meeting of the National Academy of Sciences in Schenectady on October 12. The GE scientist said that as a result of widespread seeding the previous winter and spring in California, Arizona, and New Mexico—only some of which had been conducted by Project Cirrus—silver iodide particles may have drifted hundreds of miles eastward to produce rain in the Mississippi and Ohio Valleys. Conversely, he added, overseeding might have prevented rain from falling where it was most wanted. "Thus," Langmuir said, "the drought this year in the Southwest, and the heavy rains in the Southeast, may have resulted from changes in the widespread (synoptic) weather conditions induced by the seedings in the Southwest."

His remarks startled the press. The *Schenectady Gazette* declared in a front-page article that Langmuir's tale "has changed the face of the whole United States weather map in the past 10 months." GE's hometown newspaper also noted an important claim overlooked in most other accounts: that Project Cirrus's own experiments had apparently produced "a striking seven-day rainfall cycle over areas stretching thousands of miles from New Mexico, so marked that week after week a given locality would get its heaviest rain on, say, Tuesday, its least on Saturday." That is, by seeding clouds with silver iodide smoke near Albuquerque, the project had regularly produced rain five days later in the East, a phenomenon known as periodicity. "It is thought that some effect of Project Cirrus was felt clear across the Atlantic as far as Scotland," science author Bruce Bliven later wrote. Bernard Vonnegut would warn in November that the development "may pour additional trouble upon the heads of politicians in this country and even make for a new wet-dry issue in international diplomacy."

Periodicity sparked fierce debate among meteorologists that never entirely died. The US Weather Bureau didn't believe Langmuir's findings. At the Schenectady meeting, the great GE scientist read a wire from Chief Francis Reichelderfer, which stated that the conditions observed by Project Cirrus had resulted from "major anomalies" of nature and not from silver iodide smoke. Langmuir rejected the assertion and added that despite modern tools the Weather Bureau's forecasting abilities hadn't improved in half a century.

The scientific spat rumbled on for years. Langmuir and his Project Cirrus team would continue running periodicity experiments until July 1951. On the opposite side, the Weather Bureau would conduct an exhaustive statistical analysis of past weather data and find evidence of naturally occurring periodicity. Still, the strong cycles of 1949–1950 that Langmuir reported were compelling. "The whole experiment was a great tragedy," an unnamed meteorologist would tell *Fortune* magazine in 1953. "If Langmuir actually influenced the weather, no one will believe him. If the periodicities were mere coincidence, nature played Langmuir a dirty trick." In the source notes to her book *The Brothers Vonnegut*, biographer Ginger Strand speculates that the speaker was Wallace Howell, or perhaps Vincent Schaefer or Bernard Vonnegut.

When this latest Langmuir-Reichelderfer dispute first flared up in October 1950, however, Howell kept mum. The dustup highlighted just how little was known at the time about weather in general and rainmaking in particular. Only four years had passed since Vince Schaefer's flight to scatter dry ice above Mount Greylock and Vonnegut's groundbreaking experiments with silver iodide. Howell himself couldn't yet know how effective New York City's rain stimulation project was or what unintended consequences might result. One reason he had wanted a contract extension was to conduct research through all four seasons. Now Langmuir was publicly warning about remarkably long-lived effects of silver iodide smoke. To some observers it looked as if New York City's seeding operation, "much vaunted as the answer to the age-old cliché that nobody does anything about the weather, is backfiring in some sort of way," the *New London Day* in Connecticut noted in an editorial. "Now that man has managed to do something about the weather, it appears that he can't do much about what he has done about it." The *New York Herald Tribune* pointedly added, "Although Dr. Langmuir didn't say so, his findings may mean that if Dr. Wallace E. Howell, New York's official rain-maker, is getting any precipitation at all, he may well also be watering a considerable portion of the Atlantic Ocean."

Howell's ground team again seeded Catskill-bound clouds on November 2, when the newspapers' attention was focused elsewhere. President Truman had escaped an assassination attempt by Puerto Rican nationalists at Blair House in Washington the day before; as two gunman attempted to storm the building, a wild shoot-out ensued, killing one assailant and one White House police officer and wounding the other gunman and two other officers. The war in Korea, meanwhile, had entered a dire new phase, with hundreds of thousands of Chinese troops pouring south across the Yalu River to rescue the North Korean army from advancing American and United Nations forces. "All hope of an early end to the Korean war was abandoned," the United Press bleakly reported.

And in New York City, with the mayoral election approaching, Democratic bosses were jittery over Vincent Impellitteri's lead in straw polls. "The only hope for Ferdinand Pecora, Democratic-Liberal candidate for Mayor, to defeat Acting Mayor Impellitteri lies in a strong party vote for him in Kings County," the *Brooklyn Eagle* reported the day before the vote. "Queens and Richmond [Staten Island] virtually have been conceded to the Acting Mayor, and Manhattan is suffering from a bad split within Tammany ranks." Propelled by a wave of populism and throw-the-rascals-out voter sentiment, "Impy" openly predicted victory and had every reason for optimism.

The November 7 election was a landslide: 1,157,000 votes for Impellitteri versus 937,000 for Pecora. Republican Corsi had fewer than 383,000, with American Labor's Ross under 150,000. (Meanwhile, in the governor's race, Republican incumbent Thomas Dewey was comfortably reelected.) While short of an absolute majority, Impellitteri's "precedent-shattering victory" made him the only New York mayor ever elected without the support of a major political party.

"This shows that the people realize they can do something about the conditions in the city," Impellitteri said. "I pledge to serve all the people of New York and it will be a government of all the people. The people will have an unbossed, uncontrolled Mayor." He didn't mention the man who would now largely control the office, construction coordinator Robert Moses. The acting, soon-to-be-actual mayor added that the resignations of some city commissioners would be "highly welcome."

New York inaugurated its 101st mayor a week later, in brilliant, chilly sunshine on the steps of city hall. Only five thousand citizens looked on, a tenth of the crowd that had watched O'Dwyer's farewell address. As he had promised, Impellitteri kept the ceremonies short. "The oath-taking required three minutes, from 1:07 to 1:10 P.M.," the *New York Times* reported, "and the Mayor took only about three minutes more for a talk in which he promised 'at all times to do my level-best to justify the confidence you have reposed in me.'"

It wasn't foremost in the mayor's mind on this happiest of days, but the city's water usage was again outpacing rainfall. The watershed reservoirs were down to 63.4 percent of capacity, which was about average for mid-November. Commissioner Carney pointed out to journalists

Mayor Vincent Impellitteri is all smiles after his upset victory in the November 1950 general election. *AP Photo / Anthony Camerano*

that extra rainfall in the watershed during the late winter and early spring had replenished the water supply before. "Mr. Carney hopes that something like this will happen again this year," the *New York Herald Tribune* reported, "but he emphasized that the city has no control over precipitation, 'Dr. Howell to the contrary, notwithstanding.'" The commissioner soon had cause to regret that innocent remark.

20

Thanksgiving

Thanksgiving was horrific. On Wednesday evening, November 22, 1950, the start of the long holiday weekend, the Long Island Rail Road had an accident. An all-electric train bound for Hempstead rammed the rear of a stopped Babylon express on elevated tracks a mile west of the Jamaica station in Richmond Hills, Queens.

"The Thanksgiving Eve crash came at 6:20 p.m., at the height of the homecoming commuter rush hour, when thousands of Long Island residents were homebound for the holiday," the *Brooklyn Eagle* reported. "Many of the victims were on their way to family gatherings for Thanksgiving Eve dinners which were never eaten." Seventy-eight people were killed outright in the telescoped passenger cars or died later of injuries. Well over three hundred were hurt, more than a dozen of them critically. Referring to the bankrupt railroad, the *New York Times* called the accident "the grimmest disaster of its ill-starred career."

Mayor Impellitteri rushed back from a holiday break in Cuba. Governor Dewey did the same from a getaway in Florida. The Republican governor soon demanded the resignations of two railroad trustees, and the independent mayor backed him. New Yorkers, meanwhile, observed the holiday as best they could. "War clouds in Korea and the immediate memory of a railroad disaster here cast their shadows yesterday over the city as its people observed a bountiful Thanksgiving Day," the *Times* reported.

Few New Yorkers imagined that the weekend could get worse. The *Eagle* alerted its readers to deteriorating weather conditions on Friday evening: "A few snow flurries tonight will herald a windy blast sweeping Eastward after setting all-time low temperatures in the Mid-West." Keen-eyed *Times* readers might have noticed a one-sentence news item buried next to the shipping news on Saturday morning: "PORT ARTHUR, Ont., Nov. 24 (Canadian Press)—A dozen freighters took refuge in Thunder Bay last night from one of the severest storms of the season on Lake Superior." Neither piece reflected the scope of the disaster that was roaring toward the city and the Catskills to compound the man-made miseries.

Born of a combination of weather events, the storm that struck on Saturday was New York City's third great tempest in a dozen years. It began as a huge bulge of Arctic air that plunged southward from Canada. Snow reached Alabama and Georgia, with freezing temperatures all the way to the Gulf of Mexico. At the same time, warm, moist air blew from the southeast. "As a result, a low pressure center started to develop late Friday in western North Carolina, first becoming apparent on our weather maps late Friday evening," the US Weather Bureau explained later. The storm roared up the Eastern Seaboard, "the most severe of its kind on record" and the worst since 1913. Unseasonal high pressure over the Atlantic blocked the low pressure air from moving quickly out to sea. "Instead, the disturbance has tended to move slowly in a great circle, subjecting large areas to a prolonged lashing by wind and snow," the bureau stated.

Where they collided, these forces released violent, almost unfathomable energy, making meteorologists' use of the military term *front* seem especially apt. Barometric readings fell nearly an inch in eighteen hours. Towns and cities to the west of the long moving line came to a standstill beneath blizzards; those on the eastern side hunkered beneath hurricane-force wind and rain.

Wintry chaos extended to Appalachia and the lower Great Lakes. People would tell their grandchildren about the storm for decades afterward. A newspaper weather map showed its path from southwest to northeast across all or portions of Kentucky, Virginia, West Virginia, Ohio, Pennsylvania, New York, and southern Ontario. "Tons of snow continued to pour down on northern Ohio tonight, transforming its big industrial cities into seemingly deserted ghost towns," the Associated Press reported from Cleveland. Twenty-nine people died before the snow stopped falling in the Buckeye State.

Twenty inches of snow fell on Weston, West Virginia. Two feet fell on Erie, Pennsylvania, topped by a layer of ice. A Corsair night fighter crashed on a farm outside Meadville, thirty miles south of Erie, killing the pilot. Three people died in a light-plane crash near St. Joseph, Michigan. Twenty-seven inches of snow buried Pittsburgh, where a fifteen-minute trolley ride took two hours, hundreds of people sheltered in downtown hotels rather than try to get home, and two hundred thousand mill and store workers didn't make it to work on Saturday morning. The Penn State–Pittsburgh college football game on Saturday was canceled, along with many others across the eastern states. Fifty thousand fans *did* attend the big Ohio State–Michigan rivalry game in Columbus, but spectators barely saw the field through swirling snow. The contest would forever be remembered at both schools as the Snow Bowl.

Temperature differences on opposite sides of the front were extreme. At midnight Friday, Pittsburghers shivered at 15 degrees Fahrenheit beneath a foot of snow. Two hundred miles east, thermometers read 52 degrees at the state capital at Harrisburg. In places where readings lingered above freezing long enough, the storm delivered only lashing rain and wind, a small blessing. The dividing lines were sharp. Dunkirk, New York, situated beside Lake Erie, got a foot of snow overnight. Buffalo, forty-five miles to the northeast, just saw snarled traffic and slippery streets—ONE SNOWSTORM THAT MISSED BUFFALO, read a headline in the local paper.

Altoona, Pennsylvania, population eighty thousand, lost all electrical power. Outside Washington, DC, lightning from a "freakish electrical storm" sparked a blast in Bethesda that blew out a wall and partially destroyed a power distribution plant, blacking out parts of suburban

Maryland. Syracuse, New York, was "beaten and bruised" on Saturday by winds that gusted to ninety miles per hour and blew pedestrians off their feet. With power failing at a local hospital there, four polio patients were told it was time to take their daily allotted hour outside their respirators; fortunately, the power company restored full power before they had to go back into the iron lungs that preserved their lives. The inland storm viciously hit New England, announcing itself as "the most violent of its kind ever recorded in the northeastern quarter of the U.S.," according to Weather Bureau meteorologist Ernest Christie. The blow rivaled the great 1938 hurricane "in violence but not in duration." The death count in New England by the end of the weekend reached sixty-four.

New York City escaped the snow when the storm arrived at 12:45 Saturday morning. The five boroughs instead endured horizontal rain and hurricane-strength winds, and then temperatures that fell like shotgunned mobsters. Conditions grew so wicked that by morning the normally placid East River beneath the Brooklyn Bridge resembled the North Sea in midwinter. A fallen light pole closed the bridge itself that afternoon. "Piers on the Brooklyn waterfront were flooded and water poured into the Brooklyn Navy Yard drydocks today as wind-driven tides in the harbor rose nine feet in five hours," the *Eagle* reported. "Waters were four feet above normal high tide at 9 a.m. and still rising. Fallen trolley lines in Brooklyn halted transportation as long as an hour."

Debris toppling from buildings killed two men in Manhattan. Twenty-one other people died across the region. High winds pushed the Empire State Building an inch and a half from vertical, forcing custodians to slow the elevators for safety. The steeple of St. Peter's Church in Brooklyn, built in the blizzard year of 1888, began disintegrating about noon and later collapsed. Floods, fires, and power failures disrupted rail and subway service. The Long Island Rail Road lost electricity to trackside signals. "As a consequence trains crept along—and in view of the Thanksgiving Eve collision at Richmond Hill . . . no one was inclined to blame the road for its extreme caution," the *Eagle* reported.

Manhattan above Fifty-Ninth Street and all of the Bronx lost power for forty minutes. A thousand large signs blew down across the city. In New Jersey, three radio stations with transmitters were knocked off the air, and two four-hundred-foot steel towers in Rutherford collapsed. A pair of valuable blue magpies at the Bronx Zoo escaped after wind smashed a plate glass window at the aviary. One hundred fifty national guardsmen were called out to help on Staten Island and in Queens. Police on Staten Island used a surplus army amphibious vehicle called a DUKW ("duck") to rescue people from block after block of flooded neighborhoods. One family passed a seven-week-old baby through a second-story window.

The Cunard Line vessel *Queen Mary* delayed sailing for twenty-four hours. The Staten Island and Brooklyn ferries stopped running across the wild waterways. About a thousand small craft tore loose from their moorings around the Bronx and City Island. The US Coast Guard, caught by surprise, later complained about tardy storm warnings, but the Weather Bureau said it couldn't have issued an earlier warning because the air masses had converged too quickly. "What was predicted was weather and what happened was calamity," the *New York Times* observed in an editorial; attributing the failure to "the slow progress of a specific science."

LaGuardia, Newark, Floyd Bennett, and Mitchel airfields all suspended operations. Despite a wind gust of ninety-four miles an hour, Idlewild Airport (JFK today) somehow stayed open. At 11:02, water from Flushing Bay breached a million-dollar, 12,000-foot-long protective dyke at LaGuardia. The fast-rushing tide inundated tarmacs and hangars, trapping fifty commercial and private planes, "brackish water swirling around their tails and lapping at the wings." An iconic TWA Constellation airliner sat submerged up to its engine nacelles in corrosive seawater. Officials at LaGuardia canceled 350 flights and ordered air traffic controllers out of their tower, fearing its collapse under the howling winds.

"By noon the city looked like a gigantic Halloween prankster had been on a spree with it," the *Times* reported. "Store windows broke, buildings collapsed, trees toppled, cornices and roofs tore loose from buildings, and power lines snaked crazily through some streets." Mayor Impellitteri declared an emergency at 1:30 that afternoon. He asked all businesses to send employees home and told residents to stay put for their own safety. His Honor got a firsthand look at conditions during a ride from the Roosevelt Hotel to city hall. "The trip, which normally might take no more than 15 minutes in heavy traffic, required 45 minutes yesterday as the Mayor's car dodged fallen signs, stanchions and chunks of buildings." The declared emergency lasted until the worst of the winds passed at 8:15 that evening. Overnight, temperatures that had been in the mid-50s plunged into the 20s.

Nearly 2⅓ inches of rain had fallen on the city, with winds near or above hurricane strength during most of it. The *Herald Tribune*

A stalled car in flooding on Cortlandt Street, Lower Manhattan, in the wake of the November 25 storm. *AP Photo / John Lindsay*

noted that unlike the Atlantic hurricanes of 1938 and 1944, this storm had come from the land. "On the other hand, victims, relief workers, repairmen, and others did not have to face the prospect of freezing cold when the wind and rain of those other years had subsided."

Moving north, the storm struck the Catskills later in the day than New York City. The mountainous terrain magnified the effects of the wind and especially the rain. "There was a severe wind all day Friday and Saturday, accompanied by a downpour," the *Catskill Mountain News* reported from Margaretville. "Late Saturday afternoon streams began to rise, the mountainsides were torrents, water rushing down the sides of the hills in all directions." At five o'clock streets were still passable; by six, the Fair Street bridge and Bridge Street were both "raging torrents." The flood devastated Margaretville's little downtown and washed out century-old bridges. A seventy-two-ton gasoline tank washed away from nearby Arkville and floated onto Route 28, where it snagged on telephone lines. Washouts marooned two diesel locomotives from the New York Central Railroad line between Fleischmanns and Highmount. Even the mountaintops were perilous, with the Belleayre Mountain Ski Center east of Margaretville damaged by water racing down the slope.

The *Daily Freeman* initially thought conditions were better in Kingston beside the Hudson River. "The Kingston Police Department reported one piece of Christmas decoration down on Fair street, but other than that the windstorm last night and early this morning did little damage," the paper reported Saturday afternoon. Conditions soon worsened, however. "Esopus Creek went on a rampage when two dams gave way at the height of the storm Saturday night," the newspaper reported on Monday. One of the dams was a forty-foot earth-and-concrete structure at the resort village of Pine Hill, which in failing released the waters of Funcrest Lake. "Bridges were swept away, railroad tracks and roads were washed out, homes were damaged, and several families had to be evacuated."

The *Daily Freeman* estimated that 12 billion gallons of rainwater had sluiced into the big Ashokan Reservoir yet didn't overflow the spillway. "An even more striking gain was reported by the Schoharie Reservoir, where the water level raised 54 feet in one day, so fast that a contractor's shovel and other equipment was covered with water before it could be moved out," the paper added. The Schoharie gained about 17 billion gallons of water from the storm.

The catalog of misery in the Catskills mounted hourly. Buildings flooded, trees toppled, water mains broke, phone and power lines fell, chicken coops and small buildings floated downstream. "Ulster county police said the rush of water tore out eight highway bridges and affected an area of 25 square miles," the Associated Press reported. "No loss of life was reported immediately. Police estimated the damage at $1,000,000." The Big Indian–Oliverea Valley was cut off. The mountain communities of Fleischmanns, Shandaken, Phoenicia, and Mount Tremper all flooded

Effects of flooding in tiny Fleischmanns, New York, the day following the November 25 storm. *Delaware County Historical Society, photo by Bob Wyer*

under more than 4½ inches of rain. The region might have done better to be buried under two feet of snow like the Appalachians. Stunned residents who had complained about the rainy summer now set about recovering from the monstrous deluge autumn had delivered.

It wasn't apparent for a few days, but the Thanksgiving weekend storm wrecked New York City's rain stimulation project as well. Dr. Howell's meticulous gathering of data, his strength as a scientist, now proved his undoing as a public servant.

Ironically, one of his mobile crews had generated a rare bit of good publicity in the Catskills on the Tuesday before the storm. Two men had demonstrated a generator for a Water Department deputy commissioner in Kingston. A *Daily Freeman* reporter who watched the operation quipped that he couldn't see a thing for all the smoke. "It might be a coincidence, but a half hour later snow flurries fell on the main plant of The Freeman three miles away," the newspaper reported, publishing a photo of the equipment beneath a cheeky headline addressed to Howell: WHAT SAY, DOC, SNOW FOR CHRISTMAS? The quip didn't seem funny anymore.

On Monday, in the aftermath of the storm, the owner of the Rock Cut Lodge and Cottages at Mount Tremper, north of the western tip of the Ashokan Reservoir, mistakenly believed he saw one of Howell's mobile teams generating silver iodide smoke. He was outraged. "What are they trying to do," he asked, "wash us down the stream body and bones?" With his property already damaged by floods, "man-made rain added to nature's contribution would just about wreck us." The resort owner hotly asked, "When is someone coming to our aid?"

What most people didn't yet know was that one of Howell's teams had in fact seeded the Catskills-bound clouds with silver iodide smoke on Saturday, *during* the storm. The *New York Times* published a brief article about it on Tuesday, November 28. Howell had phoned from Boston to order the seeding, which began at noon Saturday from Fahnestock

State Park. Chief Engineer Clark's deputy abruptly shut the operation down at five o'clock, the *Times* reported, and scrubbed planned seeding on Sunday as well.

"It was explained yesterday that Saturday's rain-making was called off, not because of any fear that scientific processes were related to the day's violence of nature, but because the two men operating the generator were needed back at the Ashokan headwork's for emergency storm duties," the *Times* added. The *Catskill Mountain News* noted this hasty recall as well, beneath a headline that labeled it FLOOD'S BEST JOKE.

In his final report to the city, Howell later called the storm "extraordinary" and wrote that very heavy rainfall coincided "almost exactly" with the probable path of silver iodide smoke from his team's generator, both aloft and along the ground. He didn't claim, however, that this seeding actually had caused more rain than might have fallen anyway. He simply didn't know enough to say, and he hadn't had the radar equipment that might have told him.

The generators were back in operation on Wednesday, November 29, as the Schoharie Reservoir overflowed and overall watershed water levels topped 76 percent of capacity. "Apparently unabashed by last week's storms and their effects on the city's water supply, Dr. Wallace E. Howell, the city's consultant meteorologist, telephoning from Boston, sent two trailer-borne ground generators out to burn silver iodide in the Catskills yesterday," the *Daily Freeman* acidly reported in Kingston. The generators were at work again the following Sunday.

Howell's dedication and scientific rigor didn't impress anyone in the Catskills. The *Catskill Mountain News* later concluded that the November 25 flooding around Margaretville was Howell's fault. Several times the weekly publication referred to the deluge as the "rainmaker's flood"—despite the fact that the mountain village was nearly sixty miles northwest of the ground generator's seeding site on the opposite shore of the Hudson. "When will the rainmaker find himself an honorable job and let God alone?" the weekly's Pine Hill correspondent demanded. Back in the city, on the other hand, the *New York Times* found some good in the storm, noting that 40 billion gallons of water ultimately flowed into the reservoirs. "The ill wind and the rains that accompanied it on Nov. 25 wreaked plenty of havoc, but New York City's Catskill and Croton reservoirs undeniably were beneficiaries," it stated in a December 5 editorial.

Irving Langmuir further roiled the editorial waters four days later with an attention-grabbing but ill-timed declaration. The Nobel laureate suggested in Schenectady that the federal government assume the responsibility for weather modification, as it had for atomic energy. "In the amount of energy liberated," Langmuir said, "the effect of thirty milligrams of silver iodide under optimum conditions equals that of one atomic bomb."

The statement detonated in the Catskills with only slightly less force. "Did the Belleayre mountain ridge, which poured a devastating flood down both the Dry Brook and Esopus watersheds on the night of November 25, have a charge of iodide somewhat akin to an atomic bomb?" a columnist for the *Catskill Mountain News* demanded to know. New York City's experimental cloud seeding, never popular among mountain residents, had fueled a storm of a different sort.

21

Winter

New York City's water situation improved a bit during December. On Tuesday the twelfth, the anniversary of the city's low-water mark of 33.4 percent in 1949, storage levels in the Catskill and Croton reservoirs stood at 88.8 percent of capacity. The Schoharie Reservoir still overflowed. "We are where we are today because of God's goodness and the people's cooperation," Commissioner Carney said. Chief Engineer Clark added that "there is only one chance in ten that we will not fill by next June 1." It was an astounding turnaround—but in light of the recent Catskills flooding, neither man cared to attribute it to the city's rain stimulation project.

Dr. Howell tactfully kept a low profile, but his silence didn't help the commissioner. The following week, Mayor Impellitteri let it be known that he would ask for Carney's resignation in the new year. "Politics, rather than any criticism of his work, was regarded as responsible for the decision to ask for Mr. Carney's resignation from a post that traditionally has been a reward for party service," the *New York Times* reported with elegant understatement. Carney's unforgivable sin was supporting Justice Pecora in the recent election. One of the mayor's assistants conveyed the demand on December 22. The commissioner then went to city hall to hand in his resignation.

"Frank Sampson phoned me to say that you desired my resignation," Carney's letter stated. "He explained that while you felt my work as commissioner was satisfactory, there were certain political commitments which you had to fulfill. Accordingly, I am submitting my resignation to take effect at your pleasure."

Newspapers applauded the departing commissioner's service during the drought, which had been above and beyond what was generally expected of a political appointee in New York City. The *Brooklyn Eagle* characterized Carney's ouster as "purely a political move," noting that Impellitteri had pledged not to be "vindictive" toward Pecora supporters. Carney had "labored long and hard in the fight to lick the water shortage," the *New York Post* added. "Unfortunately for himself, he also contributed his bit to a recent municipal effort to lick Vincent Impellitteri. As a result the Mayor has demanded and received Carney's resignation, thereby affirming once again the ancient doctrine that political blood is thicker than water."

The day after Christmas, the mayor named as Carney's replacement Dominick F. Paduano of Queens, who was the Water Department's first deputy commissioner and a supporter of a local labor leader who had backed Impy's run for the top office. Carney wasn't the last Pecora supporter to be shown the door at city hall. The same day, the city's acting fire commissioner transferred three top officials of the Uniformed Fireman's Association back to full-time uniformed duty, "for the good of the service." The trio likewise had made the mistake of supporting Pecora.

With Stephen Carney's departure, Howell effectively was out as well. The ax didn't fall immediately, because his city contract ran until February 20, 1951, but all that remained for the Harvard meteorologist was to write his final report. Howell's cloud-seeding teams didn't conduct any more air or ground missions in the Catskills, and his name largely fell from the newspapers. An exception came in late January, when he spoke in Manhattan at a meeting of the American Meteorological Society. His topic was a different sort of precipitation: the 384,000 tons of soot the city produced annually, and which settled darkly across the landscape as far as forty miles away.

Howell returned to the news as a rainmaker on February 14. "Rain-making experiments over New York City's watershed, held in abeyance for more than two months, are expected to end officially next Tuesday," the *New York Herald Tribune* reported. Howell told journalists that his crews had been ordered to stop seeding due to "turbidity" in the watershed. "The rainmaker explained that heavy rainfall had cut new drainage channels into the reservoirs, resulting in muddy water and reduced run-off." The *New York Times* added that "the meteorologist is expected to submit a formal report soon to Commissioner Paduano contending that at least some of the present water bonanza can be credited to his scientific efforts."

Paduano, however, saw no reason to continue the city's rain stimulation experiments, unless they were "carried out on a regional basis with New York being relieved of any responsibility." The new commissioner planned to recommend to the mayor and the Board of Estimate that they let what the *Brooklyn Eagle* called Howell's "$100-a-day-when-he-works contract" expire. "So far, Paduano said, the entire cloud-seeding project has cost less than $50,000."

"We're sorry to see Dr. Howell leave us, for as rainmakers go, he has been a good investment," the *Herald Tribune* commented the next day, in an editorial reprinted as far away as Wisconsin. The newspaper cited Howell's modesty and his generosity in crediting New Yorkers for their conservation efforts. His experiments were unlikely to settle the controversy over the effectiveness of cloud seeding, but the paper pointed out that the city's reservoirs stood at 99.6 percent of capacity, more than double the 44.8 percent of a year earlier. "The city fathers can, if they wish, release Dr. Howell for work in other, less green pastures, but we hope they give him a raincheck."

Howell nonetheless continued to boost the rain stimulation project. From his home in Lexington, he told the *Herald Tribune* that he would

like to continue his cloud-seeding experiments "to get more definitive
results." He added that "some of the rain" that had filled the reservoirs
had resulted from his seeding. "But he said it was not possible to tell
exactly how much of the rain he caused because there were too few
tests on which to base a statistical study."

Howell was still compiling his final report for city hall when his
contract expired on February 20. "The results are promising," he told the
United Press from his office at Harvard. "Statistics that I am reviewing
now and preparing for the overall report are encouraging and make me
believe that the experiments may have been successful." Asked directly
if he really had made rain, the meteorologist again used the cautious
language of a scientist. "We had results," Howell answered. "We are
not sure how definite they are. The experiments, I believe, will turn
out to be a success. They will be significant."

Nearly a year earlier, as the rain stimulation project was just getting
under way, Howell had arranged for help from the Weather Bureau in
evaluating his results. But the city was late in transferring the funds,
and the statistical evaluation wasn't far along when his contract expired.
Howell now hired college students per diem to help him compile
the results and finish his report. It was the best he could manage in
his changed circumstances. "He went into some detail in explaining
his evaluation methods," a fellow weather consultant would write in
1974, "which are similar to ones employed even today, but which were
advanced for the time." This contrasted with the Weather Bureau's
frequent criticism of Dr. Langmuir's statistical methods.

Howell's high-profile duties in New York City attracted the attention of Congress, where a joint Senate subcommittee was considering
three bills to bring rainmaking under federal control. Invited to testify
in mid-March, Howell told the senators that in his estimation cloud
seeding had boosted rainfall over the Catskills by 14 percent, adding
about 15 billion gallons of water to the reservoirs. He based that figure on data from 120 weather stations, from 1950 and the preceding
thirteen years. "He urged Federal control over weather experiments,"
the United Press reported, "and said 'conflicts' of local interests would

be inevitable if the whole problem was left to states and cities." In his official report to the city, Howell later cited a slightly higher overall figure of "about 17 percent," nearly all from ground operations. Since seeding accounted for the rise in total precipitation, and not every storm was stimulated, he wrote, the increase for seeded storms actually was closer to 39 percent.

It's unclear today exactly when he submitted his final report. Howell wrote many years later that the city's corporation counsel simply stuffed the pages into his attaché case and locked it. The *Herald Tribune* reported that the document "has been suppressed by the Corporation Counsel's office." Commissioner Paduano told journalists in late June 1951 that he hoped New York City would never again call on the Harvard rainmaker. "I haven't had an opportunity to study Dr. Howell's report on his work for the city, so I don't know what he did," Paduano said. "I hope the public will co-operate with us so it won't be necessary to get him back."

The first damage claim had struck city hall on February 16, 1951. Attorney Herman Gottfried asked for more than $1 million on behalf of the Catskill villages of Margaretville and Fleischmanns and more than eighty area property, business, and resort owners. Gottfried later called himself "just a little lawyer from Margaretville." This was false modesty. Before entering private practice in 1949, he had served as acting corporation counsel in charge of the New York City Board of Water Supply office in Kingston. In the years ahead, he would represent many Delaware County residents in disputes with the distant city over its massive water projects. Gottfried became a fierce antagonist, occasionally slapped down by judges for his tactics. In 1957 New York City claimed he held confidential information from his time with the Board of Water Supply and so "had no right to appear against the city by whom he was formerly employed." The attorney, in fact, did withdraw from several cases unconnected to the 1950 floods.

Gottfried pointed directly at the rain stimulation project in his 1951 claim for damages. "The claimants contend . . . that the city seeded clouds to induce rainfall last Nov. 25—the day heavy winds and driving rains lashed the Northeastern part of the county," the Associated Press reported. "The resulting rainfall, Mr. Gottfried said, created floods that damaged hotels, roads, bridges, and equipment." Gottfried laid out more for the local *Catskill Mountain News*: "It is believed that New York City authorities acted in a negligent manner in conducting these seeding operations after a sustained period of rain," he said. "Seeding caused an excessive and unnatural amount of precipitation."

The lawyer's claim was the first outer rain band of a legal hurricane—more were coming, and the legal storm would only worsen. On February 19, the day before Howell's contract with the city expired, Gottfried filed thirteen more claims totaling $120,833 for Catskill business owners and farmers. A Kingston firm also filed a claim on behalf of the town of Shandaken for $167,500. The little community charged New York City with "interfering with atmospheric conditions" and said that cloud seeding had resulted in "severe floods in the Esopus Creek, and other streams in the town of Shandaken." The complaint added that "it is the theory of the claim that the interference . . . was in the nature of a trespass."

At the start of summer 1951, five Kingston residents filed claims totaling $89,500 for damages caused by spring flooding along Esopus Creek—even though cloud seeding had ceased in early December. "The document held that the city's rainmaking experiment last year caused Ashokan Reservoir, which supplies water for the city, to overflow when the spring rains came and the snow melted," reported the *New York Times*. The various claimants included the owners of a private home, a nightclub, a garage, a frozen-food locker, and a training college for seals.

New York City quietly abandoned the rainmaking project altogether in August, giving over $11,000 in meteorological instruments to Hunter College in Manhattan. Howell had bought the gear for the mountain headquarters he never established; the Water Department now handed it over like so much unwanted war surplus. Commissioner Paduano said

Hunter's geology and geography department would put it to good use, but his own department had "no present or future intention to engage in seeding clouds." With a friendly nod toward Howell, the *Albany Knickerbocker News* doubted the word of Impellitteri's new man. "And we suspect that, should the situation recur, Mr. Paduano would change his tune or the city would have somebody else bearing the impressive title of commissioner of water supply, gas, and electricity."

Damage claims kept arriving: $148,650 from Greene County, $45,500 from the towns of Windham and Lexington in that county, $55,000 from the Funcrest Corporation, and more. As the first anniversary of the November 1950 storm approached, the total for flood-related claims hit $2,138,510. "The shoe was on the other foot yesterday for Father Knickerbocker," the *New York Times* reported. "Whereas a year ago the city was spending money in a handsome fashion for rain-making to end a drought, municipal funds are now being used to make a study to show whether the experiment was a failure," and so provide the city a defense against the damage claims. The newspaper noted later in an editorial that nearly two years earlier the city had hired Howell "to bring down rain and thus replenish the more than half empty Catskill reservoirs. Now the city has decided that sowing clouds with dry ice or crystals of silver iodide precipitated more trouble in the form of damage suits than rain."

The city controller blamed the 1950 Catskill flooding on the natural weather disturbance. While defensible both legally and meteorologically, the city's position intentionally ignored the scientific success or failure of the rain project. The controller's office paid out no damages to the upstate claimants, and the Impellitteri administration began to hope that the deadline for filing lawsuits—one year plus thirty days following the floods—might somehow pass unnoticed. It was a foolish notion. Four days before Christmas, attorney Gottfried stepped into the Law Department on the fifteenth floor of the Municipal Building in Manhattan toting a Gladstone bag. Gottfried had lost it temporarily when a bus left an upstate rest stop without him, but he had hitchhiked into the city and retrieved the bag at the terminal. He now upended it, loosing a cascade of summonses and complaints, "and as incredulous clerks looked on handed one fistful of lawsuits after another over the glass-topped counter."

Cumulative damage claims from 117 people and resorts in Margaretville, Phoenicia, Pine Hill, and elsewhere in the Catskills totaled $1.5 million. The language echoed Gottfried's earlier claims. He named as defendants Howell, former mayor O'Dwyer, former commissioner Carney, Chief Engineer Clark, three current Board of Water Supply commissioners, and their chief engineer. The action made news on the other side of the world: in Sydney, New South Wales, a newspaper calculated the claims in Australian pounds. New York City had twenty days to respond. Gottfried said that once he heard from the corporation counsel, "he would serve a note of issue setting the case down for trial in Supreme Court in New York County." A small newspaper in Missouri perfectly caught the attitude in the Catskills: "The city's rainmaker, Dr. Wallace E. Howell, was regarded not as the agent of heaven, but of the devilish machinations of the city government."

The city got the suits tossed out of the state Supreme Court in February 1952, on the technicality that Gottfried hadn't included the city's name in the title of his complaints. The attorney blamed "a mechanical error in preparing the summonses and complaints—which is understandable in view of the fact that there are approximately 120 of these actions." Justice Samuel H. Hofstadter believed the mistake was "a mere oversight" but wrote, "Though the plaintiff's plight may be unfortunate, it cannot override the jurisdictional barrier." The corporation counsel's office said afterward that Gottfried couldn't reinstate the actions because the thirteen-month statute of limitations had expired.

The city's victory was short lived. The state Supreme Court's Appellate Division ruled in June that blocking the suits for a failure to include the city's name placed "form above substance." The three-to-two opinion read: "The papers clearly and unmistakably gave notice to the city that it was a defendant in the action. This is demonstrated by the whole context of the complaint. The city was not prejudiced [by the omitted name] as it fully understood and indeed acted on such notice."

The setback was New York's second in a rainmaking case. A Supreme Court justice in March had smacked down the city for shenanigans in claiming that an attorney for Ulster County should have filed his paperwork not at a clerk's office but personally to the corporation counsel. "Photographs were offered in evidence showing a sign on the door of room 1532 of the Municipal building which read 'Serve papers here,'" the *Kingston Daily Freeman* cheerily reported. "Photographs of a similar sign inside a window were offered." These cases, too, were allowed to proceed, along with various others. Attorney John E. Egan, who had tried unsuccessfully to block the city's rainmaking experiments before they began, now sought damages for numerous claimants in Ulster County.

As always, New York City fought back long and hard. Each side appealed decisions that favored the other as the cases dragged on for years. Ulster County filed a motion in 1955 to grill city workers about their work on the rain stimulation project. The motion stated that "the plaintiff is unable to find eyewitnesses who are capable of telling exactly what employes [*sic*] of the City of New York did." One of those questioned was Howell's former team leader, John Aalto. The conflict stretched on into 1960, when the *Daily Freeman* published a long series of articles looking back at the then decade-old controversy. "Some $2,500,000 in lawsuits against the city of New York are still to be dealt with, it was learned today," the newspaper reported, "and some of these were filed long after the rainmaker flushed the clouds of alleged flood-making rains." (The equivalent sum today would be more than $20 million.)

The plaintiffs ultimately abandoned their suits, and none ever received a dime for damage caused by the Catskill floods of November 25, 1950. The outcome was unpopular but just. The claims had all arisen from anger and frustration rather than from verifiable science. Yes, Howell's men had seeded clouds shortly before the flooding. But the effect could only have been very limited and localized. Despite Dr. Langmuir's "one pellet" theory, a few hours of silver iodide smoke couldn't appreciably worsen a rare, complex weather pattern that had damaged property with snow, rain, and floods from the Great Lakes to New England and ended the lives of 226 people along the Eastern Seaboard.

The *New York Times* noted much later that flooding was always a problem in the Catskills, "where villages and hamlets dot the banks of the creeks and streams that thread through the valleys and hollows. But the hazards of this arrangement were laid bare during Tropical Storm Irene, as flooding turned Victorian homes to kindling, collapsed cement bridges, felled woods, and splattered once-verdant settings with mud and debris." Written in 2011, the description applied equally well to events sixty-one years earlier.

Though none of the claims against New York City for the 1950 flooding came to anything, many Catskill residents did benefit later from unrelated claims arising from water-related construction projects in Delaware County. "Farmers, business men, wage earners, and others upstate who lose land, trade, jobs, or esthetic values as a result of extension of New York City's water supply system are collecting handsomely," the *Times* reported in 1957. Upstate legislators had secured these liberal payouts, the newspaper explained, as the original cost of allowing the city's Board of Water Supply to "inundate villages, reroute highways, and divert the flow of streams more than 100 miles from urban kitchen faucets and lawn sprinklers."

Herman Gottfried often represented residents in these later cases. In Delaware County, the *Binghamton Press* reported in 1961, the attorney was considered "a David bringing the Goliath of the East Coast to its knees. . . . If he is unpopular anywhere, it is among a few of New York City's lawyers and officials who howl that awards to his clients are excessive." Some Catskill claimants, at least, finally held their own in disputes with the big city over water.

Epilogue

D r. **Howell** had learned a great deal while working for New York City. He expressed no bitterness over the ugly demise of the rain stimulation project and simply moved on with his career. The meteorologist vowed, however, that "it would be a long time before he got caught again in his own rain between irate farmers and resort owners," *Flying* magazine reported. "Instead, Howell went after Business."

"Everything in America had a price tag," historian Theodore Steinberg later wrote, "and now even the weather seemed to be entering into the realm of markets and commodities as weather companies . . . sprung up to capitalize on the bold new technology." Howell formed W. E. Howell Associates in Cambridge, Massachusetts, in spring 1951. By the following winter, the firm was listed among the top three rainmaking operations in the United States, along with Dr. Irving Krick's company in Denver and another in California. Howell Associates, a journalist noted, "devotes most of its attention to industries in need of water."

In November 1952 Howell again made the front page of the *Kingston Daily Freeman*, which had never been a fan. This time, however, the article read like something out of *National Geographic* or *Boy's Life*. The newspaper noted an Associated Press report that the meteorologist had recently returned from two jobs in the high Andes mountains of Peru, "where a mining company and some plantations in the valleys below wanted more mountain rainfall." His team had camped on glaciers and made snow a few times. "Later, he said his team moved to a spot on the continental divide 14,000 feet high near Trujillo, Peru, and used silver iodide sprays, causing showers every day."

By 1953 Howell had staffers working all over the world, sometimes flying the company's own small planes. No doubt one of his favorite projects was a rain stimulation program for the Aluminum Company of Canada. Any rain that spilled over usually benefited timber interests. "When our rain stimulating activities pour over into other acreage, we generally get cheers instead of lawsuits," he said.

The meteorologist lost a former colleague that year, with Stephen Carney's sudden death from a heart attack on August 18. The *New York Times* recalled the former water commissioner as the creator of Thirsty Thursdays. "Besides asking the public to forego bathing, shaving, car washing, and lawn sprinkling on that day, Mr. Carney hired a meteorologist to seed clouds over the Catskill watershed in the hope of inducing a rainfall," his obituary read. "By late spring water reserves had increased."

A month after Carney's death, New York City Mayor Vincent Impellitteri lost the Democratic primary election in a landslide to Manhattan borough president Robert Wagner. A *Brooklyn Eagle* editorial stated that "as things went along there did not seem to be enough of the new and independent attitude for which people were looking" from Impellitteri. His second try at an independent candidacy failed when a court ruled that he didn't have enough signatures on his petitions, effectively ending his unlikely career as Wagner won the general election.

Project Cirrus, meanwhile, had completed its government contract in September 1952. Bernard Vonnegut soon left General Electric to join the consultancy firm Arthur D. Little; he later went on to conduct research into thunderstorms and electrification. Vincent Schaefer also left GE, in 1954, to become director of research at the Munitalp Foundation (try spelling it backward) and continue his distinguished if unusual career. Irving Langmuir continued his outspoken ways until his death in 1957. "In addition to his rain-making researches, Dr. Langmuir developed techniques that led to brighter and cheaper electric lighting and that helped make modern broadcasting possible," the *Schenectady Gazette* remembered. The National Park Service later designated his Schenectady home a National Historic Landmark.

Howell flourished all the while. Journalists quoted him in articles on rainmaking and weather modification throughout the 1950s. *Fortune* magazine in May 1953 noted his work in Cuba. That summer, Irving Rosenthal of Palisades Park and the National Association of Amusement Parks, Pools and Beaches negotiated to hire Howell to prevent rain from dampening the customers. The offer was serious this time, but Lloyd's of London wouldn't provide insurance to protect "against legal proceedings from farmers or others who might feel that weather-making activity would cause them damage by preventing natural rainfall." A year later, *Collier's* mentioned a Howell Associates cloud-seeding operation that was followed by copious rainfall that helped to douse a forest fire in Quebec. The Associated Press delighted in reporting in 1957 that the company had hired meteorologist Charles Wetterer, "who seems aptly named for the job." (He went on to become a partner.) Howell kept up his outside interests as well, and in 1958 was elected president of the Mount Washington Observatory.

With the turn of the decade, the *Kingston Daily Freeman* again displayed a more appreciative view of rainmaking than it had during Howell's tenure with the city. "Cloud seeding is a top romance with nature and stirs almost as much interest as space flight," *Freeman* reporter Charles R. Douglas wrote in 1961, after covering the topic for years. "If it becomes the big science it purports to be, we'll have much more than ordinary changes of weather to talk about one of these near years ahead."

In fact, when drought again settled over upstate New York the following summer, upstate dairy farmers and fruit and vegetable growers east of the Hudson River turned to Howell Associates. "Dutchess and Columbia County farmers decided last week to try rainmaking, while those in Ulster apparently continue wary of the idea since the filing of lawsuits against New York City after the 1950 rainmaking venture," the *Daily Freeman* reported. The farmers paid Howell's firm $5,000 to produce silver iodide smoke for a month from twenty-five ground generators. Rainfall in the two counties afterward was 12 percent above normal. "While some areas received considerably more rain than others, the above normal average points to the apparent success of the program," the *Albany Times-Union* reported two summers later, when area farmers again hired Howell's company to stimulate rainfall.

Howell's business continued to thrive, both domestically and abroad. His son Stephen helped with seeding operations before entering medical school. He always recalled his stint in Peru, especially a party thrown by German sugar planters at a hacienda. The men all insisted that the elder Howell join them in an impromptu shooting match. Dr. Howell put aside a journal he was perusing, fired one shot, hit the bulls-eye dead center, and returned to his reading—no marksman's nightmare here.

In November 1964 the National Academy of Sciences and National Research Council issued a fifty-six-page report titled *Scientific Problems of Weather Modification*. The document hit Howell Associates much as floodwaters had devastated Catskill communities fourteen years earlier. The fatal paragraph appeared in the introduction. The unnamed authors noted that defenders of commercial rainmaking often argued that large numbers of farmers, companies, and government agencies spent big sums on cloud-modification services—and that only satisfied customers ever hired a rainmaker more than once. "There can be no answer to such an argument, based as it is on faith and on hope of economic gain," the authors wrote, "except to point out that the theories and predictions of astrology could be substantiated in a like manner."

The authors didn't directly equate rainmaking with astrology, merely noted the common arguments made for both. They did cite the "extremely pessimistic conclusions" of two recent studies of cumulus-cloud seeding, while also favorably noting one of Howell's studies and suggesting that additional studies might be needed. But the unfortunate word *astrology* was what newspapers latched onto, and it damaged Howell's business and those of his contemporaries. "This report was received with shock and amazement by those scientists who had by now become part of the weather modification community," a weather consultant recalled years later. Newspapers simply couldn't resist the story. SO RAINMAKERS ARE PHONY AFTER ALL, blared a headline in Newark, New Jersey.

Howell Associates and other companies began to crumble in the aftershocks. "Basic weather scientists challenge Howell's operational methods and evaluation techniques as not being scientific enough. The same goes for the more than 30 commercial seeders in the country," a Syracuse newspaper reported in mid-1965. "Howell's seeding contracts were cancelled shortly after the [astrology] announcement," a colleague wrote years later. "A subsequent review of Howell's operational data led to a controversial conclusion that his work had indeed produced increases in precipitation."

The meteorologist struggled to keep afloat. In August 1965 New York City faced another drought, fifteen years after the one that had made him famous. Howell offered to stimulate rainfall once more, this time for a penny per thousand gallons. "Wallace E. Howell of Howell Associates offered to seed clouds in the Catskill area, 'payment to be made solely on the basis of rainfall or streamfall as shown by a formula to be computed and applied by an impartial body such as the New York State Water Resources Center,'" the *New York Times* reported. Builder Fred C. Trump, father of future president Donald J. Trump, offered $10,000 to pay for the first billion gallons. "I'm interested just as a public service," the senior Trump declared at a press conference. "I'm willing to start the ball rolling just to get some water relief for the City of New York." The city didn't accept his offer.

Howell sold his company to EG&G in 1967, when the defense contractor was investing in oceanology and weather modification. EG&G made him a vice president, but corporate life didn't suit him. Howell left in 1971 to join the US Department of the Interior, Bureau of Reclamation, Division of Atmospheric Water Resources Research. The Denver-based bureau was heavily involved with Project Skywater, a weather-modification program funded by Congress in 1961. Howell helped to develop technology for increasing the water levels in bureau reservoirs, and "worked to remove the persistent doubt in the meteorological community about the effectiveness of weather modification."

Other government weather projects at this time used silver iodide. During the Vietnam War, the US Air Force launched Operation Popeye, whose slogan was "Make Mud, Not War." *New York Times* investigative reporter Seymour Hersh later described the project as "waging a systematic war of rain on the infiltration trails of Laos, Cambodia, North Vietnam, and South Vietnam. The intent: suppress enemy anti-missile fire, provide cover for South Vietnamese commando teams penetrating the North and hinder the movement of men and materiél from North Vietnam into the South." News of Operation Popeye horrified Bernard Vonnegut, whose early work had demonstrated the seeding potential for silver iodide.

From 1962 to 1983 the Weather Bureau and Defense Department jointly conducted an ambitious project called Project Stormfury. Researchers investigated the possibility of reducing the size of hurricanes and damage they caused through silver iodide seeding. Results were disappointing. The researchers concluded that seeding didn't work because hurricanes have too few supercooled water droplets and too much ice. They also couldn't distinguish the results of seeding operations from what the storms would have done anyway. "It was disappointing to concede we couldn't really do it," a Stormfury pilot said later. "But the storms were so big, and we were so small."

On the other hand, none of the hurricanes that Project Stormfury seeded suddenly changed course toward the US mainland, either. After the Hurricane King controversy in 1947, the "idea of trying to throttle hurricanes [with dry ice] fell into public disfavor," writes weather historian Cynthia Barnett. After Stormfury "failed to control hurricanes with silver iodide seeding, the idea fell into public disfavor, too." Considerable resources are devoted today to early warning, giving coastal residents time to flee. But Barnett adds that in the aftermath of catastrophic hurricane seasons such as 2017, "the risks of trying to subdue a hurricane may come to seem more acceptable too."

Howell continued happily working on government research in Denver. He also earned six patents during the 1970s, including two for snowmaking apparatuses. He published his long-suppressed final report to New York City on his Catskills work in the *Journal of*

Weather Modification in 1981, thirty years after submitting it. After retiring from the Bureau of Reclamation in 1984, Howell continued working as a consultant. He remained physically active and was still skiing at age seventy-two. The Weather Modification Association presented him with a lifetime-achievement award in 1986. Howell's name continued to pop up occasionally in articles about rainmaking or weather modification.

As the youngest among them, Howell outlived every other major figure from the 1950 rain stimulation project. Langmuir, Schaefer, and Vonnegut of General Electric; O'Dwyer, Impellitteri, Carney, and Clark of New York City; and Chief Reichelderfer of the US Weather Bureau all passed away before him. Wallace Howell died on June 12, 1999, at age eighty-four. He was survived by his wife, five children, and thirteen grandchildren.

Meteorologist Wally Howell in his seventies. *Courtesy of Howell family*

The science of weather forecasting and modeling made remarkable advances during the seven decades that followed the New York drought of 1949–1950. Widespread National Weather Service warnings about a dangerous "bomb cyclone" along the Eastern Seaboard in January 2018 would have validated Chief Reichelderfer's faith in his old bureau (now renamed) and amazed Dr. Langmuir with their timeliness and accuracy.

Weather modification, however, saw few similar advances. The question of whether New York City's audacious rainmaking project was effective still has no clear answer. Scientists, meteorologists, and researchers have offered contradictory figures and opinions ever since. An indication that silver iodide, at least, might hold promise is that companies and government agencies still use it in cloud-seeding programs in the United States and around the globe. Researchers in recent years have far surpassed Bernard Vonnegut's early experiments, using airborne radar and lasers to show that under certain conditions "the chain of events that are hypothesized to occur when you add silver iodide to a cloud do indeed occur." The use of silver iodide "keeps coming back because of the demand for water, especially the dire straits of the arid Southwest," an atmospheric research scientist said in 2009. "It's always been a cheap way to add additional water. . . . If you're trying to increase rain or snowfall for the water supply, a 10 percent addition could do a lot."

The National Research Council issued a report in 2003 on a two-year, $1.1 million cloud-seeding project in Colorado. This same council had helped to prepare the critical "astrology" report in 1964, and it found that not a lot had changed during the nearly forty intervening years. "Evaluation methodologies vary but in general do not provide convincing scientific evidence for either success or failure," the Colorado report stated. The authors suggested more long-range research, which might take decades to produce meaningful results. "In other words," a pair of science professors commented in the *Los Angeles Times*, "we won't know whether weather modification can help solve our problems until 2030, 2040, or beyond."

The next major study to gain widespread attention was the state-funded, $13 million Wyoming Weather Modification Pilot Project (WWMPP). It comprised 154 experiments across three Wyoming mountain ranges during the winters of 2008–2014. An executive summary indicated that silver iodide seeding from ground generators could increase snowfall by 5 to 15 percent, with little harmful effect on the environment from the silver iodide used in the testing. A WWMPP scientist later added, "In the core of the silver iodide plumes, we may see the snowfall rate double or more, according to the model." The Wyoming figures overlapped those that Howell quoted in his final report in 1951. The WWMPP results weren't conclusive, however, and silver toxicity from any large-scale testing still worries many scientists and environmentalists.

As recently as spring 2016 the Los Angeles County Department of Public Works used ground generators and "flare trees that poke above the scrub like dull gray cacti" to seed clouds with silver iodide. The county estimated it had increased rainfall from seeded clouds by about 15 percent—again, roughly matching Howell's estimate from 1951. In 2017 a research team dubbed SNOWIE (an acronym for "Seeded and Natural Orographic Wintertime clouds: the Idaho Experiment") seeded clouds with silver iodide from aircraft in the Snake River Basin. "To get the complete picture, they're gonna need a bigger box—a supercomputer," *Popular Science* magazine reported. The SNOWIE team had access to one called Cheyenne, among the fastest computers in the world. They planned to use it to see how well physical observations in the field matched predictions. "And based on how well they do or don't, the SNOWIE team and other scientists can then tweak the predictors to better see which weather is the most fertile for modification." The Desert Research Institute in Nevada, meanwhile, experimented with drones to seed clouds with silver iodide flares, in hopes of reducing the dangers of seeding from aircraft in unstable conditions.

Other countries experimented as well. In 2017 rainmaking projects were underway or planned for Taiwan, India, Australia, Jordan, Malaysia, Vietnam, and the United Arab Emirates, as well as several

American states. China especially has always tried to control its weather. During the nineteenth century, following a long drought near Beijing, an angry emperor officially exiled one of the dragons popularly believed to control rain. The country's current approach is distinctly more modern. Once the Communist government came to power in 1949, *Vanity Fair* later declared, "the rain dragons were slaughtered and banished from thought."

"The country's army of rainmakers," China's National Meteorological Bureau announced in 2006, "uses rockets, artillery, and aircraft to sow rain-inducing chemicals at times of need." The government was ready to fire silver iodide in artillery shells to prevent rain from disrupting the Summer Olympic Games in Beijing in 2008. The following winter, Chinese scientists fired hundreds of cigarette-sized packets of silver iodide into clouds above drought-stricken northern provinces. After a snowfall heavy enough to close a dozen highways around the capital, a senior engineer called the seeding "a procedure that made the snow a lot heavier."

China didn't stop there. The Natural Development and Reform Commission announced a 1.15 billion yuan (about US$168 million) rainmaking project in 2017 for the country's dry northwestern provinces. "The NDRC approved the budget to buy four new planes, upgrade eight existing aircraft, develop 897 rocket launch devices, and connect 1,856 devices to digital control systems," Hong Kong's *South China Morning Post* reported. "The whole project will take three years."

None of the modern developments would have surprised Wally Howell. He likely would have been disappointed, though, that scientists and meteorologists are still arguing about the worthiness of cloud seeding and weather modification. The reasons for such disagreements would've been familiar, too. The intermingling of science and technology with politics and the law, spiced with the continuing conflict between the rights of urban and rural populations, still impedes progress.

Weather also remains massively difficult to understand and predict, as demonstrated by the "spaghetti tracks" of multiple destructive hurricanes in the Caribbean and the Gulf of Mexico and along the Atlantic Seaboard in 2017. The butterfly flaps its wings, but where does the resulting storm develop? A municipality seeds a cloud, but how much of the rain that falls on thirsty hills would have fallen anyway? Howell didn't know in 1950. Generations later, meteorologists and researchers don't understand a great deal more. Research continues. As Bernard Vonnegut's brother Kurt once wrote, "So it goes."

Wallace Howell stood in the forefront of his field at the midpoint of the American Century, an era defined by science and a popular belief that with science nearly anything was possible. Much has changed since, not entirely for the better. In noting the Gotham rainmaker's passing in 1999, *Weatherwise* magazine recalled "Howell's Snow" in New York City, his pioneering rainmaking flights during the 1949–1950 drought, and his subsequent work around the world. "Although the saying goes 'Everyone talks about the weather, but no one does anything about it,' there are always exceptions," an editor wrote. "And Wallace E. Howell was an extraordinary exception."

Acknowledgments

Thank you to my agent, Kelli Christiansen, for her advice, support, and unfailing good cheer; to Dr. Stephen B. Howell, for generously sharing memories of his father, and to my friend Sara Shopkow, for her research in the New York City archives. Thanks, also, to meteorologist Mike Davis, of WBNS-10TV in Columbus, Ohio, for reviewing various scientific details, and to developmental editor Devon Freeny of Chicago Review Press, for keeping a complex narrative on track. Any remaining errors are mine alone.

Notes

Introduction

"celebrated rain making efforts": Henry Fountain, "Wallace E. Howell, 84, Dies; Famed Rainmaker in Drought," *New York Times*, July 6, 1999.

"to whom nothing": Morris, *Manhattan '45*, 7.

1. Drought

"Colonial roads, stone walls": Merrill Folsom, "Low Water Bares Colonial Relics Once Hidden in Kensico Reservoir," *New York Times*, January 14, 1949.

"warmest year": "Mercury Drops to Season's Low as Warmest Year Here Nears End," *New York Times*, December 31, 1949.

"At Ithaca": "Air Mass Change Needed to Provide Relief from Heat," *Kingston Daily Freeman*, June 14, 1949.

"Every fisherman and camper": "A Parched Land" (editorial), *New York Times*, June 16, 1949.

"disgustingly damp": "Rain (1/100 Inch) Falls on City; No Relief Seen for Thirsty Crops," *New York Times*, June 20, 1949.

"stood by their fences": Steinbeck, *Grapes of Wrath*, 3.

"Drought ordinarily is something": "The City Man and the Lack of Rain," *Brooklyn Eagle*, June 28, 1949.

"A sudden, short-lived": "Thunder Shower Cools Arid City," *Brooklyn Eagle*, July 6, 1949.

"The first few drops": "Rain" (editorial), *New York Times*, July 10, 1949.

"An archivist, writing": "Great Water Famine of 1910 Makes '49 Drought 'All Wet,'" *Mount Vernon Daily Argus*, August 1, 1949.

"*He said he had prepared*": "Thunder Shower Cools Arid City," *Brooklyn Eagle*.

"*teetering on the brink*": Leo Turner, "Drought May Soon Force Topers to Drink Their Liquor Straight," *Brooklyn Eagle*, October 26, 1949.

"*This would divide*": "New York City Is Facing Serious Water Shortage," *Kingston Daily Freeman*, December 3, 1949.

"*on the ground*": "Rain Expected Today to Lift City Supply," *Brooklyn Eagle*, December 4, 1949.

"*ghost city, with no power*": "Warns of Disaster in Water Shortage," *Brooklyn Eagle*, December 7, 1949.

"*normally an estuary*": Orton, "Hudson River or Estuary?"

"*one of the most notable*": Flinn, "The World's Greatest Aqueduct," 707.

"*some time in 1952*": Quoted in "City Rainmaking HQS. Set Up at Bennett Field," *Brooklyn Eagle*, March 15, 1950. Carney's estimate of the completion date would prove optimistic and inaccurate.

"*one of the wonders*": "Driving the Shandaken Tunnel," *Earth Mover*, 11.

"*As a result*": Emily S. Rueb, illustrations by Josh Cochran, "How New York Gets Its Water," *New York Times*, March 30, 2016.

"*It is therefore evident*": Groopman, "New York Dependable Supplies," 828.

"*1,000-square-mile watershed*": "That Drought," *New York Times*, December 18, 1949. Many other newspapers in New York and elsewhere used the same phrase during 1949–1950.

"*The large size*": Van Burkalow, "The Geography of New York City's Water Supply," 375.

"*Sixty thousand acres*": Silverman and Silver, *Catskills*, 195.

"*derived their name*": Irving, "Catskill Mountains," 408.

"*Since the new Catskills*": Stradling, *Making Mountains*, 185.

"*Throughout his later career*": "Former Mayor O'Dwyer Dead; Prosecuted Murder, Inc., Gang," *New York Times*, November 25, 1964.

"*As mayor, O'Dwyer lacked*": Arthur Everett, "Immigrant to Mayor, 'Bill' O'Dwyer Dies," *Albany Knickerbocker News*, November 25, 1964.

"*either by voluntary co-operation*": "Warns of Disaster," *Brooklyn Eagle*.

"*After that it was just*": "Showing the Way to the City," *Rockaway Beach Wave*, December 22, 1949.

"*two or three villages*": "Mayors Claim Ample Water for Villages," *Chatham Courier*, December 8, 1949.

"*We must conserve water*": "Warns of Disaster," *Brooklyn Eagle*.

"*The prelate ordered*": "Spellman Asks New York Catholics to Pray for Rain, End Drought," *Dunkirk Evening Observer*, December 10, 1949.

"*Give us, we pray*": Harold Faber, "Rains Upstate Fail to Halt Lowering of Water for City," *New York Times*, December 12, 1949.

"*absolute need to avoid*": "Catholics to Offer Prayers for Rain," *Yonkers Herald Statesman*, December 10, 1949.

"*If you want to be able*": Faber, "Rains Upstate."

"*My husband walked out*": "East Side, West Side, All Around N.Y., Taps Are Off," *Tonawonda Evening News*, December 16, 1949.

"*Stop Wasting Water*": Sam Falk, "Observed Waterless Friday," *New York Times*, December 17, 1949.

"*It is entirely possible*": Bob Lanigan, "Some TV Personalities and a Few Anecdotes," *Brooklyn Eagle*, March 6, 1953.

"*When New Yorkers hang up*": "Manhattan Water Outlook Brightens, Heavy Drain Halted," *Binghamton Press*, December 24, 1949. Chief Engineer Clark's hope would ultimately prove accurate. City officials and forecasters couldn't know it at the time, but the reading of 33.4 percent of capacity on December 12, 1949, had marked the bottom of the drought. Levels wouldn't dip so low again.

"*If winter brings heavy rains*": "New York Wells Fast Running Dry; Big City Faces Drought Until 1952," *Elmira Star-Gazette*, December 8, 1949.

2. Snow

"*the kind of child*": "Battled Years over Atoms of Air," *Brooklyn Eagle*, March 21, 1926.

pack a blue serge suit: Strand, *Brothers Vonnegut*, 23.

"*was exploring mountains*": Schaefer, "Irving Langmuir, Versatile Scientist," 483.

"*uncanny ability to turn*": Hugh Taylor, "Irving Langmuir," 167.

"*raised lighting levels*": "Retired, but Will Continue Work," *Schenectady Gazette*, January 3, 1950.

"*His outstanding accomplishments*": "Langmuir Has New Theory of Energy," *Brooklyn Eagle*, May 2, 1920.

"*the greatest development*": "Irving Langmuir—Creator of the Super-Tube," *Popular Science Monthly*, October 1922.

"*sitting at the transmitting key*": "Marconi's Mystery Signals Not from Mars but Radio Station at Schenectady," *Brooklyn Eagle*, June 27, 1922.

"*an entirely new branch*": "Retired, but Will Continue," *Schenectady Gazette*.

"*who continually embarks*": "Personalities in Science—Dr. Irving Langmuir," *Scientific American*, September 1937.

"*Train yourselves*": Langmuir, "Find Yourself This Summer."

"*approached a problem*": Schaefer, "Irving Langmuir, Versatile Scientist," 483.

"*The smoke screen*": "G.E. Smoke Machines Mask Monty's Drive," *Schenectady Gazette*, March 24, 1945.

"*Langmuir the Indomitable*": Horace R. Byers, "History of Weather Modification," in Hess, *Weather and Climate Modification*, 15.

"*In his work and in his diversions*": "Schaefer Wins Civic Service Award," *Schenectady Gazette*, January 22, 1941.

"*Too frequently the awarding*": "One in a Thousand" (editorial), *Schenectady Gazette*, June 8, 1948.

"*It didn't take long*": Don Rittner, preface to Schaefer, *Serendipity in Science*, xv.

"*Stories like this are possible*": "What It Takes to Make a G-E Scientist" (General Electric advertisement), *Concordiensis*, February 17, 1950.

"*perhaps superior to*": Vincent J. Schaefer, "Irving Langmuir, Man of Many Interests," *Science*, May 28, 1958.

"*Well-trained men*": Bruce Lambert, "Vincent J. Schaefer, 87, Is Dead; Chemist Who First Seeded Clouds," *New York Times*, July 28, 1993.

"*snowflake scientist*": Steven M. Spencer, "The Man Who Can Make It Rain," *Saturday Evening Post*, October 25, 1947.

"*fortunate accident*": Schaefer, "Irving Langmuir, Versatile Scientist," 483.

"*Now I know how*": Strand, *Brothers Vonnegut*, 53.

"*It is a wonderful experiment*": Langmuir, *Atmospheric Phenomena*, 161.

"*featuring Vincent Schaefer*": General Electric, "Dr. Vincent Schaefer Snow-Making Demonstration," www.youtube.com /watch?v=2D5s2FlA_5k.

"*by scattering small fragments*": Schaefer, "Production of Ice Crystals."

3. Mount Greylock

"*I looked to the rear*": Vincent J. Schaefer, "Cloud-Bombing Flight to Make Snow Described by Scientist Who Did It," *Albany Times-Union*, November 15, 1946.

"*a radical modification*": G. B. Lal, "GE Experts Turn Cloud into Snow," *Albany Times-Union*, November 14, 1946.

"*the rapidity with which*": "'Project Cirrus,'" *General Electric Review*, 12.

"*This is history*": Strand, *Brothers Vonnegut*, 57.

"*That simple cloud seeding*": Henderson, "Vincent J. Schaefer: A Remembrance," 163.

"*a snowstorm that never took place*": Drake, "Rainmakers Are All Wet."

"*In a five-hour flight*": "Three-Mile Cloud Made into Snow by Dry Ice Dropped from Plane," *New York Times*, November 15, 1946.

"*Man-made snow*": General Electric press release, November 16, 1946, facsimile in Schaefer, *Serendipity in Science*, 405.

"*near magical experiment*": "Man-Made Snowstorm," *Life*.

"*sensational conquest over nature*": Lal, "GE Experts."

"*In any event, why should*": "Dry Ice Makes Cloud Yield Snow" (editorial), *Rome Sentinel*, November 16, 1946.

"*carried a hint that*": Howard W. Blakeslee, "Scientists Creating First Man-Made Snowstorm Amazed to See Cloud Proceed to Change Shape," *Yonkers Herald Statesman*, December 7, 1946.

"*While the seeding group*": "'Project Cirrus,'" *General Electric Review*, 12.

"*the subsequent storm was a big one*": Schaefer, *Serendipity in Science*, 142. Schaefer recalled the fourth test flight taking place on Christmas Eve, but GE's official report placed it on December 20.

"*continued to skywrite*": Spencer, "Man Who Can Make It Rain."

"*I have produced some snow*": "Dr. Schaefer of GE Makes Real Snow," *Syracuse Herald-Journal*, November 30, 1948.

"*Dry ice could make*": Byers, "History of Weather Modification," 13.

"*they rode the breezes*": Spencer, "Man Who Can Make It Rain."

"*While the work has not yet*": "Snow Made in a New Way," *New York Times*, January 31, 1947.

"*As explained by Dr. Vonnegut*": "Scientists Discovery May Make Possible Modification of Winters Weather by Cutting Thru Clouds," *Schenectady Gazette*, October 20, 1947.

"*In one minute, the generator*": Dr. Bernard Vonnegut, "Influencing the Weather Advanced by Scientists," *Albany Times-Union*, November 30, 1947.

"*Fiercely-burning charcoal*": General Electric press release, October 28, 1948, facsimile in Schaefer, *Serendipity in Science*, 381.

4. Possibilities

"a plea for divine guidance": Thomas P. Ronan, "Mayor Gives Oath to 58 at Inaugural for Second Term," *New York Times*, January 3, 1950.

"His renomination came only": Warren Moscow, "Naming of O'Dwyer as Envoy to Mexico Is Held Eminent," *New York Times*, August 15, 1950.

"My doctor told me": "Former Mayor O'Dwyer Dead," *New York Times*.

"It would be hard to imagine": "Langmuir's Advice," *Schenectady Gazette*, January 27, 1950.

"At the Astor on the 25th": Walter Winchell, Broadway, *Albany Times-Union*, January 23, 1950.

"New York public sentiment": Charles G. Bennett, "New York Water Supply: A Long-Range Problem," *New York Times*, December 18, 1949.

"Water starved New York City": "City Granted Right to Tap Hudson River," *Brooklyn Eagle*, January 19, 1950.

"the safety, health and welfare": "State Hears City Bid to Tap Hudson; Poughkeepsie Loses Fight to Block Plan," *New York Times*, January 11, 1950.

"Rapping the city": "State Tells City It May Tap Hudson," *Long Island Star-Journal*, January 19, 1950.

"will not solve the problem": "State Water Control Board Grants NYC Right to Pump Water from Hudson," *Albany Times-Union*, January 20, 1950.

"the Hudson River work": Charles G. Bennett, "Slight Water Gain for 'Dry Thursday' Far Short of Goal," *New York Times*, January 20, 1950.

"positive indications of cloud": "'Rain-Making' Held of No Importance," *New York Times*, January 26, 1949.

"You can't make rain": "Can Man Make It Rain? Top Scientists Disagree," *Albany Times-Union*, January 26, 1949.

"self-propagating storms": "'Rain-Making' Held," *New York Times*.

"When Dr. Reichelderfer": John W. Finney, "No. 1 Weatherman Will Retire Soon," *New York Times*, April 22, 1963.

"fascinating game": "Meet Top U.S. Weather Man," *New York Sun*, November 12, 1943.

"unbenevolent features": "'Rain-Making' Held," *New York Times*.

"I believe that it is impossible": "Seeks Method for Diversion of Rain and Big Snowfalls," *Ogdensburg Advance-News*, January 8, 1950.

"It is easier to make": William L. Laurence, "Seeded Rain Noted 2,000 Miles Away," *New York Times*, October 13, 1950.

"The Weather Bureau has not": F. W. Reichelderfer, "Letters to the Times: Methods to Induce Rain," *New York Times*, March 27, 1950.

"No one knows the answer": "Can Man Make It Rain?," *Albany Times-Union*.

"dramatic play-by-play": "Langmuir Gives Play-by-Play Account of Man-Made Rain Storm in New Mexico," *Schenectady Gazette*, January 26, 1950.

"Dr. Langmuir holds": Howard Blakeslee, "Silver Iodide Smoke Launches Cloudburst over New Mexico," *Oakland Tribune*, January 26, 1950.

"highly probable": "Keep That Faucet Dry," *Yonkers Herald Statesman*, January 26, 1950.

"They ought to hire": "Langmuir's Advice," *Schenectady Gazette*.

"We can't go into": "White Rainmaker Won't Lift a Finger, Either; Clouds Aren't Right," *Tonawonda Evening News*, December 9, 1949.

"I've never made a drop": William Lowenberg, "Scientist Urges Global Weather Control by UN," *Albany Times-Union*, September 9, 1950.

"future economic benefits": "Scientist Predicts Boon to Air Safety—Sees Planes Battling Storms with Projectiles," *Elmira Star-Gazette*, August 19, 1949.

"Dr. Langmuir is no": "Rain-Making for New York," *New York Times*, January 30, 1950.

"New York City turned seriously": "Gotham to Try 'Rainmaker' as Water Crisis Grows; Dr. Langmuir Is Called In," *Albany Times-Union*, February 11, 1950.

"Mr. Carney evidently": Richard J. Lewis, "Dr. Langmuir Favors Test of Rain-Making to Aid N.Y.," *Albany Times-Union*, February 13, 1950.

"I don't want to leave": "City Water Chief and Scientist Confer on Rain Making," *Brooklyn Eagle*, February 15, 1950.

"to a large extent": "Weather Control in 10 Yrs. Is Forecast by Langmuir," *Schenectady Gazette*, February 16, 1950.

"Is there an engineer": "Rainmakers to Study Feasibility of Methods," *Schenectady Gazette*, February 25, 1950.

"As always, the News Bureau": Strand, *Brothers Vonnegut*, 151.

"I am certainly going": Richard J. Lewis, "New York Official, After Talk by Langmuir, Will Urge O'Dwyer to Hire a Rain-Maker," *Albany Times-Union*, February 16, 1950.

"*a competent meteorologist*": Kalman Seigel, "Carney Moves to Hire Expert for Project in Rain-Making," *New York Times*, February 16, 1950.
"*I am very, very enthusiastic*": "Carney Touring Watershed Area—and Hoping Weather Stays Bad," *Brooklyn Eagle*, February 16, 1950.

5. Crackerjack

"*A Harvard man will plan*": H. I. Phillips, The Once Over, *Syracuse Post-Standard*, March 4, 1950.
"*The two scientists swapped*": Richard K. Winslow, "Rainmaker Takes Job in His Stride," *New York Herald Tribune*, March 26, 1950.
"*the degree of success*": "Some Control of Weather Seen Feasible," *Schenectady Gazette*, August 20, 1949.
"*You might get a call*": Howell, "Precipitation Stimulation Project," 89.
"*was big, big money*": Fountain, "Wallace E. Howell, 84, Dies."
"*Dr. Howell said last night*": Charles G. Bennett, "Expert to Survey Rain-Making Plan," *New York Times*, February 21, 1950.
"*It is a challenging experiment*": "Harvard Expert to Help N.Y. in Rain-Making Experiment," *Boston Globe*, February 21, 1950.
"*Any knowledge gained*": "Water in the Future" (editorial), *Schenectady Gazette*, February 22, 1950.
"*in most cases, reservoirs*": Aryn Baker, "Cape Town Is 90 Days Away from Running Out of Water," *Time*, January 15, 2018.
"*He indicated his belief*": "City's Reservoirs Up to 46.9% Level," *New York Times*, February 26, 1950.
"*after we get the green light*": Charles G. Bennett, "City Speeds Plan for Rain Making," *New York Times*, March 1, 1950.
"*six months to a year*": "6-Month Delay on Rain Making Test Is Likely," *New York Herald Tribune*, March 3, 1950.
"*the prospect that the natural*": Howell, "Precipitation Stimulation Project," 106.
"*worth undertaking*": "Rain-Making Test Plans Sent to City," *New York Herald Tribune*, March 5, 1950.
"*It's a shame*": "Carney Studies Hopeful Report of Rain Maker," *New York Herald Tribune*, March 7, 1950.
"*While the winds tore*": "Dodgers Go to Bat in Water Crisis," *New York Times*, March 10, 1950.

"Has there been any": "Rain-Making Step Decided for City," *New York Times*, March 14, 1950.

"the first time rainmaking": Charles G. Bennett, "Rain-Making Tests Begin Within Week; $50,000 Fund Voted," *New York Times*, March 15, 1950.

"set aside for police": "City Rainmaking HQS.," *Brooklyn Eagle*.

"makes him the highest-paid": "Cloud Physicist," *New Yorker*, April 8, 1950.

"By watching for": Bennett, "Rain-Making Tests Begin."

"I am sure that": "City Rainmaking HQS.," *Brooklyn Eagle*.

"The rain-making experiment": Waldemar Kaempffert, "Rainmakers Get Chance to Test Their Science," *New York Times*, March 19, 1950.

"With a wild burst": "The Rain-Maker Is Hired," *New York Herald Tribune*, March 15, 1950.

"The truth is": "Rain-Maker Is Hired," *New York Herald Tribune*.

"Her thesis, as an educator": "Mrs. J. C. Howell Dead; Retired Yale Lecturer," *New York Herald Tribune*, July 12, 1951.

"I didn't find": Winslow, "Rainmaker Takes Job in His Stride."

"For his third attempt": Kahan, "In Memoriam: Wallace E. Howell," 6.

"In five years": Winslow, "Rainmaker Takes Job in His Stride."

"Shortly thereafter, Howell placed": Kahan, "In Memoriam: Wallace E. Howell," 6.

"Only since 1925": Vern Croop, "Boss Weatherman Gazes in Crystal, Sees Normal Conditions for 1950," *Utica Daily Press*, December 26, 1949.

"like many other young scientists": Saul Pett, "New York Rainmaker Finds Lexington Lawn Parched," *Boston Globe*, October 1, 1950.

"very high-powered": Stephen B. Howell, author interview, January 3, 2018.

"He is also very busy": Pett, "New York Rainmaker."

"If in the long run": Ed Creagh, "Success in Rain-Making Promises New Life in Vast Arid Areas," *Binghamton Press*, March 19, 1950.

6. Headquarters

"there is less likelihood": Howell, "Precipitation Stimulation Project," 100.

"Why shouldn't he be": "Threat of Spring Showers Worries Rainmaker," *Albany Knickerbocker News*, March 17, 1950.

"Mayor Erastus Corning was": "Wanted: Dairy Clouds," *Time*, March 27, 1950.

"*The particular interest*": "Corning Asks Control of 'Rain-Making,'" *Albany Knickerbocker News*, February 17, 1950.

"*It's a legal problem*": Frederick C. Downing, "Weathermen Take No Sides in Dispute on New York City's Rain-Making Efforts," *Albany Knickerbocker News*, February 18, 1950.

"*There's no evidence*": Bob Eunson, "Cloud Puncturing Project Starts to Moisten City," *Geneva Daily Times*, March 17, 1950.

"*It isn't likely that*": "Threat of Spring Showers," *Albany Knickerbocker News*.

"*temporary moving headquarters*": Richard K. Winslow, "Rain-Makers Start Hunt for Catskill Base," *New York Herald Tribune*, March 18, 1950.

"*A boyish looking Harvard professor*": Bob Eunson, "Professor Begins Hunt for More New York Rains," *Gloversville Leader-Republican*, March 17, 1950.

"*When I planned*": Charles G. Bennett, "Rain Makers Fail in Quarters Hunt," *New York Times*, March 18, 1950.

"*panting, puffing and blowing*": William Lowenberg, "After 3,150-Foot Climb to Overlook Mt. Summit, Scientist Tentatively Rejects Site; Search Continues Today," *Albany Times-Union*, March 18, 1950.

"*We can't do much*": Bennett, "Rain Makers Fail."

"*We would use up*": Bob Eunson, "Hunt for Site Worrying Potential Rainmakers," *Syracuse Post-Standard*, March 18, 1950.

"*Carney Slipped Here*": Winslow, "Rain-Makers Start Hunt."

"*The liner Washington*": "Sea So Rough, Goldfish Sick," *Geneva Daily Times*, March 18, 1950.

"*At one point the state car*": Robert Eunson, "Swirling Snow Fails to Deter Rainmaker's Search for Site," *Syracuse Post-Standard*, March 19, 1950.

"*athletic abilities that [had] left*": William Lowenberg Jr., "New York's Rainmaker Halts Hunt for Right Peak After Battling Snow Drifts," *Albany Times-Union*, March 19, 1950.

"*We haven't anything*": Lowenberg, "New York's Rainmaker Halts."

"*The Harvard scientist indicated*": "City Completes Search for Its Rain-Making Site," *New York Herald Tribune*, March 19, 1950.

"*They don't want to see*": William Lowenberg Jr., "Catskill Area Residents Dubious Regarding Rain-Making Proposal," *Albany Times-Union*, March 26, 1950.

"*Maybe they ought*": "Rainmaker Ends Hunt for Project Site," *Rochester Democrat and Chronicle*, March 19, 1950.

"*Isn't the weather*": Robert Eunson, "New York City's 'Rain-Maker' Seeks Catskill Spot for Test," *Geneva Daily Times*, March 18, 1950.

"*two or three airplanes*": "Airborne Picket Line May Enter Made-Rain Wrangle," *Kingston Daily Freeman*, March 18, 1950.

"*doubtful proposition*": Alexander Feinberg, "Clark Casts Doubt on Artificial Rain," *New York Times*, March 20, 1950.

"*but for the water shortage*": Winslow, "Rainmaker Takes Job in His Stride."

"*There've been reports*": Alvin Davis, "Artificial Rain a Reality—Quantity Is the Problem," *New York Post*, March 17, 1950.

"*We like the way*": "Forecaster Likes N.Y. Rain Project," *Harvard Crimson*, March 22, 1950.

"*one of the most responsible*": Steinberg, "Cloudbusting in Fulton County," 224.

"*I still hope something*": "Rain Site Hunt Will Continue," *Albany Times-Union*, March 20, 1950.

"*It was additionally reported*": Feinberg, "Clark Casts Doubt."

"*At normal consumption*": "Rain-Makers End Search for Site," *New York Times*, March 19, 1950. The *Times* had estimated a 106-day supply with less water in the reservoirs on February 26. The March 19 article didn't explain the discrepancy, which may have involved different measurements of "present consumption" on the two dates.

7. Hurricane King

"*research study of cloud particles*": "'Project Cirrus,'" *General Electric Review*, 13.

"*a vast weather research*": "Army and GE Join to 'Make Weather,'" *New York Times*, March 14, 1947.

"*If the Army could find*": Harper, *Make It Rain*, 57.

"*confined strictly to laboratory work*": Havens, *History of Project Cirrus*, 13.

"*Yet the Army*": "Destroying a Hurricane," *New York Times*, September 10, 1947.

"*Effects of the seeding*": "Storm Nears 'Buster' Test with Dry Ice," *Schenectady Gazette*, September 12, 1947.

"*Secrecy shrouded the plans*": "'Hurricane Buster' Plans Still Secret," *Schenectady Gazette*, October 11, 1947.

"*small freakish hurricane*": "Freak Hurricane Floods Miami Area, Crops Hit," *Rochester Democrat and Chronicle*, October 13, 1947.

"*Army and Navy hurricane*": "Floods Ruin Florida Crops," *Detroit Free Press*, October 13, 1947.

"*There aren't enough planes*": "Freak Hurricane Floods," *Rochester Democrat and Chronicle*.

"*The fliers waited until*": "'Operation Cirrus' Rated Success, Results Secret," *Schenectady Gazette*, October 14, 1947.

"*Even if the experiment*": "Florida Hurricane Sweeps Out to Sea, Leaving 2 Dead; Damage Will Total Millions," *Schenectady Gazette*, October 13, 1947.

"*urgent and much simpler operations*": Havens, *History of Project Cirrus*, 63.

HURRICANE DISSIPATES AFTER: "Hurricane Dissipates After Raking Savannah," *Schenectady Gazette*, October 16, 1947.

"*surged to super-hurricane*": E. V. W. Jones, "Maybe Dry Ice 'Busted' That October Storm," *Daytona Beach Morning Herald*, October 31, 1947.

"*Frankly, we don't know*": Larry Murray, "Hurricane Study Only Begun, Says Schaefer," *Schenectady Gazette*, October 17, 1947.

"*Only the Army, Navy*": "Storm Observer Lists Peculiarities of Recent Hurricane in Caribbean," *Schenectady Gazette*, October 31, 1947.

"*At no time, was any*": James L. Kilgallen, "Weather Control Far in Future, GE Expert Says," *Albany Times-Union*, September 4, 1948.

"*The result of this*": "Freak Hurricane Floods," *Rochester Democrat and Chronicle*.

"*There was not one chance*": Bill Richardson, "Langmuir Links '52 Floods to Rainmaking," *Niagara Falls Gazette*, April 28, 1955.

"*could lead to modification*": "Dr. Langmuir Urges Action," *Albany Times-Union*, August 25, 1955.

"*Therefore there is every*": "Cloud Seeding Held Little Aid in Hurricane," *New York Post*, August 7, 1956.

"*This is the kind of thing*": "Langmuir Theory Questioned: Expert Doubts Seeding Affected 1947 Hurricane," *Albany Knickerbocker News*, August 7, 1956.

8. The Goose

"simplicity itself": Charles G. Bennett, "Rain-Making Plane Satisfies O'Dwyer," *New York Times*, March 21, 1950.

"As soon as they warn": Alvin Davis, "Will 'Operation Rainfall' Succeed? 4 Scientists Have 4 Different Views," *New York Post*, March 16, 1950.

"We are offering": Richard K. Winslow, "Pilots Briefed for Early Test in Rain Making," *New York Herald Tribune*, March 21, 1950.

"a fast-growing child": Margaret C. Smith, Washington and You, *Troy Times Record*, January 10, 1950.

"This is certainly": "Cloud Physicist," *New Yorker*.

"so full of rain": "Rain, Snow, Winds Come with Spring," *New York Times*, March 22, 1950."

"wide vistas to the west": Richard K. Winslow, "Catskill Rain Sites Rejected; New Headquarters to Be Sought," *New York Herald Tribune*, March 22, 1950.

"On Aug. 3, at 5:20": "Rain Made in Britain with Ice for First Time," *New York Times*, August 14, 1949.

"prepared to move in": "Showers Due While Experts Hunt Rain," *Long Island Star-Journal*, March 22, 1950.

"detect and record more accurately": "Radar Will Check City's Rain Test," *New York Times*, March 22, 1950.

"But atmospheric conditions": Eunson, "Professor Begins Hunt."

"A scheduled rain-making": "Rainfall Halts N.Y. Experiment," *Dunkirk Evening Observer*, March 23, 1950.

"but refuses to go into": "Shortsighted" (editorial), *Schenectady Gazette*, March 29, 1950.

"voodoo of a fairly base grade": Robert C. Ruark, "Rain-Making Is Nothing New," *Buffalo Courier-Express*, March 25, 1950.

"strip-teasing a cloud": H. I. Phillips, The Once Over, *Rochester Democrat and Chronicle*, March 20, 1950.

"Now if this comes to pass": "Says Prayers Brought Rain, Bills Dewey," *Binghamton Press*, March 28, 1950.

"rain day, that is": David C. Whitney, "Today Is R-Day and It Will Be 'Pellets Away' in Catskill Mountains; Dry Ice Planes Ready," *Dunkirk Evening Observer*, March 28, 1950.

"*It must be used*": Paul Crowell, "Supply of Dry Ice Ready for Clouds," *New York Times*, March 26, 1950.

"*the new menace*": "With the Sky Police," *Popular Mechanics*.

"*ex-GIs with combat*": Herbert Mitgang, "Beat of Wings: New York's Air Police Take Rainmaking as Just Another Phase of Their Duty," *New York Times*, May 21, 1950.

"*rushed about getting*": "Rain Holds Up Rain-Making Flight, Then Seeders Find No Cloud to Ice," *New York Times*, March 29, 1950.

"*You can dump dry ice*": "City Launches First Rain-Making Effort," *Brooklyn Eagle*.

"*Many a morning*": Jim McGarry, "Ex-Pilot Sees Military Use for Artificial Snow," *Schenectady Gazette*, November 15, 1946. The article posed possible aviation applications for Schaefer's snow research.

"*The Fifteenth's chief adversary*": Tillman, *Forgotten Fifteenth*, 149.

"*like a cartoon whale*": Strand, *Brothers Vonnegut*, 157.

"*an almost architectural design*": Jones, "Wild Goose Chase," 148.

"*City planes carry*": Mitgang, "Beat of Wings."

"*and the plan was to land*": "City Launches First," *Brooklyn Eagle*.

"*The rainmakers made it clear*": "N.Y. 'Rain Plane' Flying High to Catskills—with High Hopes," *Albany Knickerbocker News*, March 28, 1950.

"*They were: Lack of clouds*": Robert Eunson, "Eunson Misses Rainmaking Fiasco, but Is Successful in Finding 'Natural' Rain," *Gloversville and Johnstown Morning Herald*, March 29, 1950. Eunson didn't mention potential difficulties with airborne picketers over Grossinger's nearby resort.

"*It doesn't look as though*": "N.Y. 'Rain Plane,'" *Albany Knickerbocker News*.

"*We were held to*": "Rain Holds Up," *New York Times*.

"*but the results would*": "Rainmakers Seek Fat, Billowy Clouds to Seed," *Corning Evening Leader*, March 29, 1950.

"*a whole lot of airplane*": Jones, "Wild Goose Chase," 149.

"*At latest reports no one*": "Mother Nature Fools Rain Makers, for Free," *Long Island Star-Journal*, March 29, 1950.

"*We didn't see a good*": Richard K. Winslow, "Clouds Elude Rain-Maker Plane After Fog Here Delays Take-Off," *New York Herald Tribune*, March 29, 1950.

"What we learned will make": "Rain Holds Up," *New York Times*.

"You just don't talk": Eunson, "Eunson Misses."

CATSKILL SKIES TOO BLUE: "Catskill Skies Too Blue, They Couldn't Find a Cloud," *Rochester Democrat and Chronicle*, March 29, 1950.

CLOUDS ARE TOO SHY: "Clouds Are Too Shy for N.Y. Rainmakers," *Tonawanda Evening News*, March 29, 1950.

"Ironically, rainfall since": "Cloud Dearth over Watershed Stymies Efforts of Rainmakers," *Syracuse Post-Standard*, March 28, 1950.

"Don't worry about": Eunson, "Eunson Misses."

"We got a wonderful view": "New York's Rainmaker Fails in Hunt for Clouds," *Rochester Democrat and Chronicle*, March 29, 1950.

9. Who Owns the Clouds?

"Once man takes control": A. M. Low, "Scientist Predicts World Weather Control," *Albany Times-Union*, August 28, 1949.

"The snowflakes all evaporated": "Man-Made Snowstorm," *Life*.

"a very worrisome hazard": Havens, *History of Project Cirrus*, 13.

"has disassociated itself legally": "Weather or Not?," *Time*.

The Associated Press had pointed out: "Gotham to Try 'Rainmaker'," *Albany Times-Union*.

"Operating in the same": "Weather or Not?," *Time*.

"would rub out a lot": "Gotham to Try 'Rainmaker'," *Albany Times-Union*.

"I am only interested": "Dr. Langmuir Favors Test," *Albany Times-Union*.

"In New York City": "Possible Rain-Making for New York City Explored," *Schenectady Gazette*, February 16, 1950.

"Before tinkering with nature": "Area Residents Concerned," *Kingston Daily Freeman*, February 20, 1950.

"What would happen to": "Will City Rainmakers Cause Floods in Mountains?," *Catskill Mountain News*, February 22, 1950.

"The law department in effect": "Chance for Lawyers" (editorial), *Schenectady Gazette*, February 17, 1950.

"We're looking into": "Solution to Water Shortage?," *Life*.

"a horrendous lawyer's": "Wanted: Dairy Clouds," *Time*.

"Is there any valid": "Haymaking and Rainmaking" (editorial), *New York Times*, August 27, 1950.

"Common law provides no": Anderson and Howell, "Should the U.S. Government Control the Rain Makers?"

"The problem of water diversion": "New York to Try Rain-Making" (editorial), *Buffalo Evening News*, March 15, 1950.

"The article goes clear back": "Rainmaking Produces Odd Legal Problems," *Syracuse Herald-Journal*, December 3, 1948.

"In most jurisdictions": "Who Owns the Clouds?," *Stanford Law Review*, 63.

"It would be nonsense": Brooks, "Legal Aspects of Rainmaking," 115–116.

"The legal angles": "Every Man a Rain Maker?," *Changing Times*.

"new field of law": Quoted in "Little Drops of Water . . ." (editorial), *Buffalo Courier-Express*, March 6, 1950.

"cloud commissions": Claire Cox, "Rain-Making Attempts May Require Arbitration Board," *Cedar Rapids Gazette*, March 19, 1950.

"All we can do now": Booton Herndon, "Legal Problems of the Rainmakers," *American Weekly*, May 21, 1950.

"The trade groups claimed": Richard K. Winslow, "City, Sued, Goes Ahead with Test of Rain Making," *New York Herald Tribune*, March 23, 1950.

"huge and false propaganda": "Charge Made N.Y. Falsifies Water Supply Problem," *Kingston Daily Freeman*, March 22, 1950.

"grave danger": Charles G. Bennett, "Suit Seeks to Block Rain-Making, but City Will Not Halt Its Tests," *New York Times*, March 23, 1950.

"Somebody doesn't want": Bennett, "Suit Seeks to Block."

"At the risk of being": "Proof of Damage" (editorial), *Schenectady Gazette*, March 27, 1950.

"legal waterspout": "City Will Tap Up-State Clouds Despite Legal Deluge," *Brooklyn Eagle*, March 23, 1950.

"In the absence of": Winslow, "City, Sued."

"Water Czar Stephen J. Carney": "Hot, Humid Weather Due This Weekend," *Long Island Star-Journal*, June 24, 1950.

"Look, our guests come": "Weather or Not?," *Time*.

"The relief which plaintiff": "For Public Good" (editorial), *Schenectady Gazette*, May 15, 1950.

"up to the Legislature": "State Gives Wary Okay to City Rain-Making Plan," *New York Post*, March 8, 1950.

"to make a comprehensive study": Bennett, "Suit Seeks to Block."

"the usual operations": "Every Man a Rain Maker?," *Changing Times*.

10. Achilles' Heel

"The high temperature": "Early Thaw Boosts N.Y. Water, Rainmaker Idle," *Albany Times-Union*, March 30, 1950.

"Despite the fact": "Honest Showers Send Mountain Rainmakers Home," *Catskill Mountain News*, March 31, 1950.

"No one is a miracle man": "It's Dry Day, Carney Warns Water Users," *Long Island Star-Journal*, March 30, 1950.

"a Dry Thursday": Richard K. Winslow, "Reservoirs Up 7,303,000,000 Gallons in a Day," *New York Herald Tribune*, March 30, 1950.

"a complete failure": Richard K. Winslow, "City Gets Offer of 3 DC-3s Free in Rain Making," *New York Herald Tribune*, April 1, 1950.

"with deepest gratitude": Charles G. Bennett, "3 DC-3's, Radar Join Rain-Making Fleet," *New York Times*, April 1, 1950.

"The Sperry planes": Winslow, "City Gets Offer."

"Bucket brigade": Jeanne Tomey, "Census Tabulators Discover Brooklyn Is Full of Surprises," *Brooklyn Eagle*, April 2, 1950.

"There was just too much": Fendall Yerxa, "Fear of Creating Floods Delays City's Attempt at Rain Making," *New York Herald Tribune*, April 2, 1950.

115 days of water: "Full Reservoirs by June Unlikely," *New York Times*, April 4, 1950.

"Gordon's job calls for": "Gordon Sails into Headlines Due to Work at Reservoir," *Kingston Daily Freeman*, June 6, 1950.

"He telephones his reports": "Man Ignored 41 Years Makes Big News Now," *Wakefield News*, September 15, 1950.

"that the need for voluntary": "Melting of Snow Helps Reservoirs," *New York Times*, April 3, 1950.

"This thing just happened": "Watershed Rise Fails to Avert City's Shortage," *New York Herald Tribune*, April 3, 1950.

"I'm taking the wraps": "Rain-Maker Told to Get Up and Do Stuff," *Brooklyn Eagle*, April 4, 1950.

"a locality not shown": Richard K. Winslow, "Rain-Makers Select 'Ideal' Catskills Site," *New York Herald Tribune*, April 5, 1950.

"It is planned to have": "Rain-Maker Is Set for New Try Today," *New York Times*, April 5, 1950.

"*an excellent outlook*": "Howell Selects Lakewood Site for His Headquarters," *Kingston Daily Freeman*, April 4, 1950.

a bold X: "Where City Will Set Up Rain Headquarters," *New York Times*, April 5, 1950.

"*in a race with a towering layer*": "N.Y. Rainmaker Proposes, but God, as Usual, Disposes," *Tonawanda Evening News*, April 5, 1950.

"*the Civil Aeronautics Administration*": "Rain-Maker Tries It Again; Turns Back, Blaming Radio," *Brooklyn Eagle*, April 5, 1950.

"*It seemed pointless*": Richàrd K. Winslow, "Static Halts 2d Rain-Maker Try; Supply Rises, Today Is Dry Day," *New York Herald Tribune*, April 6, 1950.

"*succeeding better than man*": "Man Still Running 2nd to Nature in Filling Up Our Water Reservoirs," *New York Post*, April 6, 1950.

"*Shall we have expended*": John A. Heffernan, Heffernan Says, *Brooklyn Eagle*, April 6, 1950.

"*explained how to get*": Winslow, "Static Halts."

"*There are some old*": "Says Many Trees Drink Too Much," *Kingston Daily Freeman*, May 26, 1950.

"*worthless and unproductive trees*": "Forest Program for Water Urged," *New York Times*, May 26, 1950.

11. Mystified City

"*Meanwhile, the city*": Charles G. Bennett, "Catskill Freshnets Halt Rain-Making," *New York Times*, April 8, 1950.

"*last year's drought*": "Watershed Snow Ceasing Run-Off," *New York Times*, April 11, 1950."

"*It is because we realize*": Bennett, "Catskill Freshnets."

"*Nature is providing*": "Attorney Says Overflow Proves Shortage Is 'Hoax,'" *Kingston Daily Freeman*, April 7, 1950.

"*Thousands of Brooklynites*": "Watershed Snow," *New York Times*.

"*stormy petrel*": Frank J. Taylor, "They Make Rain."

"*appreciable results*": "Dr. Krick Calls City Rain Making Haphazard and Waste of Money," *New York Herald Tribune*, April 13, 1950.

"*A truck tows a small trailer*": "Rainmaking Show Is Put on Uptown," *Kingston Daily Freeman*, November 21, 1950.

"*we got going as quickly*": "Dr. Krick Calls," *New York Herald Tribune*.

"perhaps to try again": "Rainmaker Takes Off," *Kingston Daily Freeman*, April 13, 1950.

"Even so, he said": Richard K. Winslow, "City Seeds Clouds for First Time; Snow Falls, but Cause Not Clear," *New York Herald Tribune*, April 14, 1950.

"triggerable cloud formation": "Carney Happy, No Matter Whose Snow," *Long Island Star-Journal*, April 15, 1950.

"It would be completely impossible": "NY Rainmaker 'Seeds' Clouds over Watershed," *Schenectady Gazette*, April 14, 1950.

"It wasn't snowing": Charles G. Bennett, "Dry-Iced Clouds Yield Snow but Expert Takes No Credit," *New York Times*, April 14, 1950.

"pattern of records": "NY Rainmaker 'Seeds,'" *Schenectady Gazette*.

"Whatever snow we caused": Winslow, "City Seeds Clouds."

"baby blizzard": "Snowstorm Not Work of Rainmaker," *Syracuse Herald-Journal*, April 14, 1950.

"A mystified New York": Charles G. Bennett, "'Howell's Snow' Irks Some but City Calls It Fine Stuff," *New York Times*, April 15, 1950.

"The Winter-in-mid-April": "Snow, Record Cold Hit City," *Brooklyn Eagle*, April 14, 1950.

"Merry Christmas!": "Winter's Return Brings City 1.5-Inch April Snow," *Yonkers Herald Statesman*, April 14, 1950.

"Look at that snow": "Stoneham Calls upon Weatherman for New Schedule," *Rochester Democrat and Chronicle*, April 15, 1950.

"Stoneham could set up": Vince Johnson, Once Over Lightly, *Pittsburgh Post-Gazette*, April 18, 1950.

"This is written": "Pssst! Know What! We've Got Snow!," *Suffolk County News*, April 14, 1950.

"The mysterious snow storm": "Nature-or-Man-Made Snowstorm Ices Highways, Slows Traffic," *Long Island Star-Journal*, April 14, 1950.

"Your damned shenanigans": "Our Rain Is His Snow, Newtown 'Mayor' Says," *New York Times*, April 23, 1950.

"New Yorkers were calling it": "Gotham Gets Snow for Sure, but Did Rainmakers Do It?," *Binghamton Press*, April 14, 1950.

"Irate motorists muttered": "The Weather: Who Made It Snow?," *Newsweek*, April 24, 1950.

"There was one man": "Carney Happy," *Long Island Star-Journal*.

"DID I DO THIS?": "Spring's on a Binge; Cold Records Tumble," *St. Petersburg Times*, April 15, 1950.

"convinced that the airplane": "Believes Snowfall Due in Part to Cloud-Seeding," *Brooklyn Eagle*, April 16, 1950.

"We may have helped": George W. Cornell, "Clouds 'Sowed' by Rainmaker; All East Reaps!," *Utica Observer-Dispatch*, April 14, 1950.

"I wouldn't say the cloud-seeding": "Gotham Rainmaker Seeks Spigot to Stop Downpour," *Saratoga Saratogian*, April 14, 1950.

"It might cause a storm": Richard K. Winslow, "City's 5-Hour Surprise Snowfall Was Not the Rainmakers' Doing," *New York Herald Tribune*, April 15, 1950.

"The thoughtlessness of those people": "Snow Complaints" (editorial), *Schenectady Gazette*, April 18, 1950.

"We think God caused": "April's Disputed Showers" (editorial), *Pittsburgh Post-Gazette*, April 15, 1950.

"We are sitting tight": "Man Proposes, God Disposes: Manhattan Orders Rain but Heavens Send Snow," *Rochester Democrat and Chronicle*, April 15, 1950.

12. Jupiter Pluvius

"pale face": "Your Village All Done, Leave It, Hopi Rainmakers Advise N. Y.," *Tonawanda News*, December 9, 1949.

"And it is said that": Plutarch, *Plutarch's Lives*, 9:521.

"When numerous observers": Powers, *War and the Weather*, 3–4, 76, 85.

"Senator Stanford, for example": "Artificial Rainmaking," *New York Evening Post*, April 25, 1891.

"Not only did we": "The Rainmakers," *Fort Worth Gazette*, August 28, 1891.

"Most of the material": "To Show Off at Midland," *Fort Worth Gazette*, August 28, 1891.

"It will be remembered": "An Original Rain-Maker," *Helena Daily Independent*, December 20, 1891.

"The first thing I have": Alexander Macfarlane, "The Rainmakers," *Austin Weekly Statesman*, December 8, 1892.

"no satisfactory results": "A Booming Failure," *Austin Weekly Statesman*, December 8, 1892.

"I never tried": "John Jacob Astor's Latest Invention," *New York Tribune,* March 27, 1893.

"A dispatch from Fort Scott": "The Rainmakers," *Prescott Arizona Weekly Journal-Miner,* August 17, 1892.

"The fact that storms": C. S. Boyd, "Rain Is Not Produced by Cannonading— Is One of Nature's Most Gigantic Operations," *Washington Herald,* October 24, 1915.

"found his confreres in science": "Oliver Lodge Dies; Noted Scientist, 89," *New York Times,* August 23, 1940.

"By placing a copper rod": "Scientist to Control Weather," *Guthrie Daily Leader,* January 27, 1914.

"If we want rain": "To Make Our Own Weather and Control Our Crops by Girdling the Earth with a Copper Belt," *Richmond Times-Dispatch,* April 19, 1914.

"That's too deep": "Scientist to Control Weather," *Guthrie Daily Leader.*

"the ghost of Sir Oliver Lodge": "'Sir Oliver Lodge's Ghost' Here on Holiday; Has Eye on $10,000 Offered for Real Spook," *New York Times,* September 3, 1940.

"He was self-assured": Brimner, *The Rain Wizard,* 22, 37.

"the skies, in an attempt": "Rainmaker Will Invade Valley," *Los Angeles Herald,* October 28, 1906.

"It is therefore apparent": *Toledo Blade,* reprinted in "Weather Chief Raps Hatfield," *Los Angeles Herald,* March 7, 1905.

"those who are credulous": "Rain Making" (editorial), *New York Times,* May 15, 1905.

"What his chemicals are": Tales of Gotham and Other Cities, *Chicago Eagle,* April 1, 1916.

"Just hold your horses": Tuthill, "Hatfield the Rainmaker," 108–109.

"The city attorney further avers": "A Dry Town's Troubles" (editorial), *Tacoma Times,* January 26, 1916.

"Hatfield has been a rainmaker": "Are You Superstitious?" (editorial), *Seattle Star,* February 15, 1921.

"a brilliant eccentric": "Archibald Montgomery Low," British Interplanetary Society, accessed August 15, 2018, www.bis-space.com/what-we-do/the -british-interplanetary-society/history/a-m-low.

"Dr. Low admitted": "Says He's Invented Seeing by Wire," *New York Times*, May 29, 1914.

"Perhaps fortunately for man": Low, "Scientist Predicts World Weather Control."

"Nash's rainmaker is not": Hischak, *100 Greatest American Plays*, 251.

"the phony rainmaker": N. Richard Nash, "Path of a 'Rainmaker,'" *New York Times*, December 9, 1956.

"News of his death": "Rainmaker Dies at 82," *New York Times*, April 15, 1958.

"Needless to say": Henry Fountain, "The Science of Rain-Making Is Still Patchy," *New York Times*, October 19, 2003.

13. Combined Operations

"one of the most extravagant": Stuart W. Little, "Rain Makers Attacking Clouds from Land and Air in New Test," *New York Herald Tribune*, April 20, 1950.

"Under starless skies": "Dual Attack Set by City on Clouds," *New York Times*, April 20, 1950.

"Millions of silver iodide": Little, "Rain Makers Attacking."

"highly improbable": "5 Billion-Gallon Fall in Catskills; Some of It May Be Rain Maker's," *New York Herald Tribune*, April 21, 1950.

"The weather thoroughly fouled": "Rain Keeps Rain-Maker from Making the Same," *Brooklyn Eagle*, April 20, 1950.

"It appeared that officials": "5 Billion-Gallon Fall," *New York Herald Tribune*.

"I wish we could": "Heavy Rains Drench Watersheds; Howell Again Disclaims Credit," *New York Times*, April 21, 1950.

"It was nature's rain": "Heavy Rains Drench," *New York Times*.

"They may have eked": "5 Billion-Gallon Fall," *New York Herald Tribune*.

"Whether Howell intended": "Howell Goes to Work in Rain Storm," *Long Island Star-Journal*, April 20, 1950.

"I fervently hope": "Hopes Are Slim for 100% Store of Water June 1," *New York Herald Tribune*, April 25, 1950.

"instruments for readings": "City's Rainmakers Spraying Clouds Again with Silver Iodide Smoke over Watershed," *New York Times*, April 26, 1950.

"From a quick, casual check": "Howell Takes Some Credit for Watershed Rain," *New York Herald Tribune*, April 27, 1950.

"There was rain over": "Rain Was Heavier over Seeded Area, Says Dr. Howell," *Brooklyn Eagle*, April 27, 1950.

"Speaking at a regular meeting": "Speaker Says Flood Conditions Are Not Rainmaker's Wish," *Kingston Daily Freeman*, April 28, 1950.

"Later, a large group": Harold Faber, "Loyalty Marchers Ignore Rain, Cold," *New York Times*, April 30, 1950.

"The experiment started": "No Claim Made by Rainmakers," *Kingston Daily Freeman*, May 1, 1950.

"Dr. Howell kept": "3-Hour Rain Falls in 16-Hour Seeding," *New York Times*, May 1, 1950.

reported they went: "Crew Tries Again to Smoke Out Rain," *New York Times*, May 2, 1950; "Rain Crews Try Again, Supply Rises, Use Dips," *New York Herald Tribune*, May 2, 1950.

14. Marksman's Nightmare

"You see, my pitchers": Bob Cooke, Another Viewpoint, *New York Herald Tribune*, April 30, 1950.

"While you have been reticent": "Amusement Park Hit, Rain Maker Gets Offer to Quit," *Syracuse Post-Standard*, May 3, 1950.

"research on the constitution": "Offer to Stop Making Rain Is Turned Down," *North Adams Transcript*, May 9, 1950.

"Dr. Howell cannot operate": "Sunmakers Plan to Foil Rainmaker," *Schenectady Gazette*, May 8, 1950.

"By broadcasting sound": "'Saboteurs' Get $500 to Kibosh $100 Rainmaker," *Schenectady Gazette*, May 23, 1950.

"a sort of tug of war": "Rain Stopping Opportunity" (editorial), *New London Day*, May 4, 1950.

"But, Schaefer thinks": "Lightning Storms May Be Stopped by Human Means," *Troy Record*, May 6, 1950.

"a convenient high point": "Rainmakers Again Seed Upstate Area," *New York Times*, May 7, 1950.

"since there was virtually": "Smoking of Clouds Coaxes Some Rain," *New York Times*, May 8, 1950.

"that rare gem of a day": "Today's the Day: Warm and Sunny (So the Man Says)," *Brooklyn Eagle*, May 7, 1950.

"*a suggestion has been made*": "Hospital Fair Promises Host of Attractions Memorial Day," *Rhinebeck Gazette*, May 11, 1950.

"*Dr. Howell makes no pretension*": Leslie Hanscom, "At End of Drought It'll Rain, He Says!," *Brooklyn Eagle*, May 12, 1950.

"*The mobile generations*": "Iodide Squirts Clouds, Clouds Return Squirt," *Binghamton Press*, May 16, 1950.

"*cold, misty, and foggy*": "Chill and Mist Back for One-Day Stand, Say Weather Men," *Brooklyn Eagle*, May 18, 1950.

"*I am pretty darn sure*": "Rain-Maker Says Bad Weather in City Is Nature's and Not His," *New York Herald Tribune*, May 18, 1950.

"*we do know certainly*": "Howell Says God Alone Can Make Murky Weather," *Kingston Daily Freeman*, May 18, 1950.

"*I wish I knew*": John Randolph, "How Much of This Rain Is Coincidence?," *Geneva Daily Times*, May 23, 1950.

"*rain-making enthusiasts*": "Cold Rain Adds Billion Gallons to Reservoirs," *New York Herald Tribune*, May 20, 1950.

"*appeared to be the result*": "City Rainmaker Seeds Clouds, Showers Fall," *New York Herald Tribune*, May 21, 1950.

"*The plane has greater range*": "Clouds Belabored by Air and by Land," *New York Herald Tribune*, May 24, 1950.

"*Asked if he had caused*": "Catskills' Clouds Hit Top and Bottom," *New York Times*, May 24, 1950.

15. Cloud Pirates

"*Five minutes later*": "Nick the Rain Maker," *Life*.

"*Nick Gregovitch, his neighbors*": "Respectability Achieved" (editorial), *Arizona Republic*, September 5, 1947.

"*He was not cloud-cuckoo-land-crazy*": "Science: Whose Rain?," *Time*, December 22, 1947.

"*We plan to make rain*": "Who Owns the Rain?," *New York Post*, December 9, 1947.

"*California wants to hog*": "Rancher's Rain Claim Defied," *San Bernardino Daily Sun*, December 13, 1947.

"*milk our clouds*": Jack Goodman, "Mountain States: 'Cloudbusting' in Nevada Is Fought by Utah," *New York Times*, January 18, 1948.

"One can easily visualize": "Rain-Making Scientist Warns 'Cloud-Pirating' Can Lead to Trouble," *Binghamton Press*, January 8, 1948.

"As a result there was": Senate Committee on Commerce, *Weather Modification*, 48.

"has a right to the free": "'Rainmaker' Threatened," *New York Times*, August 14, 1947.

"T'other day Chickasha": Arthur Edson, "Experts Throw Cold Water on Rampaging Rain-Makers," *Amarillo Daily News*, September 17, 1947.

"Actual flying time is": C. A. Johnson, "Pilot Sells Showers," *Binghamton Press*, December 17, 1947.

"Some of the correspondents": "Rain-Maker Blamed for Sydney's Wet Summer," *Sydney Morning Herald*, January 17, 1948.

"shoots silver iodide after": "Monticello Airman Says Snow, Two Rains Were His; Used Iodide," *Kingston Daily Freeman*, May 4, 1950.

"The action came after": "Research in Rain-Making Launched by New Mexico," *Schenectady Gazette*, May 29, 1950.

"many of them woefully ignorant": "Science: Too Much Rainmaking," *Time*, June 12, 1950.

"Many farmers, ranchers": "Rainmaker Licenses Seen by GE Weather Scientist," *Schenectady Gazette*, August 26, 1950.

"sprouted like mushrooms": "Time for Controls?" (editorial), *Schenectady Gazette*, August 30, 1950.

"I'm a good cloud": "Show of Hands," *The Dick Van Dyke Show*, season 4, episode 28, aired April 14, 1965.

16. Weather Headaches

"Dr. Howell's activities": "Dry Day Failure but Supplies Gain," *New York Times*, May 27, 1950.

"I feel that it's a sacrilege": "Calls Bombing Clouds Sacrilege," *Kingston Daily Freeman*, May 27, 1950.

"The east basin can still": "Ashokan Is More Than 99 Per Cent; Rainmaker Idle," *Kingston Daily Freeman*, May 29, 1950.

"With the permission": Ed Sinclair, "Dodgers, Yankees Hosts Today to Phils, Red Sox in Twin Bills," *New York Herald Tribune*, May 30, 1950.

"Thanks to Dr. Wallace": Bob Cooke, "One-Hitter Scored at Belmont;
    4 Winners for Atkinson," *New York Herald Tribune*, June 3, 1950.

"far better off than": Charles G. Bennett, "Public Saved Enough Water to
    Cover Manhattan 12 Feet," *New York Times*, June 1, 1950.

"Still," the paper added: "55 Billion Gallons of Water Saved by Public in
    5 Months," *Brooklyn Eagle*, June 1, 1950.

"Anyone thinking of using": Charles G. Bennett, "Pool and Garden Water
    Ban Eased for 30-Day Trial Starting June 15," *New York Times*,
    June 2, 1950.

"Oddly enough, the month": "Water-Saving Efforts Bring Easing of Crisis"
    (editorial), *Brooklyn Eagle*, June 3, 1950.

"popped an aspirin": "Clouds, Showers Forecast Today," *Brooklyn Eagle*,
    June 4, 1950.

"We're not coming down": "Water Supply Up a Billion Gallons," *New York
    Times*, June 4, 1950.

"an almost empty supply": It's a Strange World, *Binghamton Press*, June 5,
    1950.

"Even if the reservoirs": Fendall Yerxa, "Ashokan Spills Over, but Croton
    Supply Is Short," *New York Herald Tribune*, June 4, 1950.

"personal cloud-seeding flight": "Water Up, Near Year-Ago Level; Usage 25%
    Under '49 Level," *New York Herald Tribune*, June 18, 1950.

four of the June storms: Howell, "Precipitation Stimulation Project,"
    96–98.

"I wonder if it's theirs": Untitled cartoon, *New Yorker*, May 13, 1950.

"His Eminence said it": "Reservoirs Rising in Thunderstorms," *New York
    Times*, June 12, 1950.

"denied, however, that Howell": "Too Much Water for Water Men," *Lowell
    Sun*, June 15, 1950.

"Despite superficial appearances": "City Bars 'Holiday' for Rain-Making,"
    New York Times, June 14, 1950.

"saddest of the hate-Howell groups": Ernie Hill, "Resorts Hit Rain-Making
    'Experiment,'" *Binghamton Press*, June 28, 1950.

"squirmed but bravely carried on": "Catskill Officials Just Loathe Howell,"
    New York Times, June 23, 1950.

"The difference between those": "Daily Water Use Shows Big Saving," *New
    York Times*, June 27, 1950.

"We are only permitting": "Water Bans Eased on Pools, Showers," *New York Times*, June 30, 1950.

"It has been offered": "Catskill System Still Overflowing," *New York Times*, June 6, 1950.

"SEOUL, Korea, Sunday": "War Is Declared by North Koreans; Fighting on Border," *New York Times*, June 25, 1950.

17. Summertime

"sheer slaughter": "845 Holiday Deaths Set All-Time Record," *Brooklyn Eagle*, July 5, 1950.

"many leaving their weekend": "Holiday Traffic Flows Smoothly in Westchester," *Yonkers Herald Statesman*, July 5, 1950.

"Rains that approached": "1.63-Inch Rainfall Drenches the City," *New York Times*, July 11, 1950.

"The heaviest July rainfall": "July Rainfall Is Heaviest in 50 Years," *Long Island Star-Journal*, July 15, 1950.

"Even though we have announced": "Downpour Raises Reservoir Levels," *New York Times*, July 12, 1950.

"Did you do it": Pett, "New York Rainmaker."

"public nuisance": "Rain-Making Suit Threatened," *New York Times*, July 11, 1950.

"may be subjected": "Supervisors Kick on Rain-Makers," *Gloversville Leader-Republican*, July 11, 1950.

demanding that the city "desist": "2d County Protests Rain-Making," *New York Times*, July 15, 1950.

"rains which have fallen": George E. Yaeger, "Rain Making Protested," *New York Times*, July 27, 1950.

threats to take potshots: "Did Rainmaker Bring Wet Season to the Catskills?," *Catskill Mountain News*, August 4, 1950.

"I know some people": "Farmers Air Rainmaker Views; Some Would 'Shoot on Sight,'" *Yonkers Herald Statesman*, August 23, 1950.

"The Catskills, the greatest": Mountain Dew, *Catskill Mountain News*, July 21, 1950.

"The result has been": "Dr. Howell Finally Admits Efforts Have Made Rain," *Catskill Mountain News*, August 11, 1950.

"*Attention Dr. Howell*": "Forecast for Whole Month: Fair and Warm; Why Don't You Let Us Alone, Rainman?," *Kingston Daily Freeman*, July 19, 1950.

"*as portentous and pervading*": Paul C. Friedlander, "Clouds over the Catskills," *New York Times*, August 13, 1950.

"*Since the A-bomb*": Drake, "Rainmakers Are All Wet."

"*RAINMAKERS ARE ALL WET*": "'Rainmakers Are All Wet'—So Says Egg Farmer," *Schenectady Gazette*, March 10, 1948.

"*The use of a single pellet*": Langmuir, "Control of Precipitation," 36.

EASY TO MAKE RAIN: "Easy to Make Rain, Langmuir Asserts," *New York Times*, July 16, 1950.

"*In those years*": Pro Bono Publico, "Rainfall in New York City," *New York Times*, August 7, 1950.

"*There's a persnickety*": Billy Rose, Pitching Horseshoes, *Utica Observer-Dispatch*, August 2, 1950.

"*water use in the generally hot*": "August Water Use to Decide Curbs," *New York Times*, August 2, 1950.

"*Dr. Wallace E. Howell, the city's*": "Dr. Howell Yields; Yes, He Made Rain," *New York Times*, August 8, 1950.

"*showed no signs*": John O'Reilly, "August Called Not as Cool as Many Believe," *New York Herald Tribune*, August 8, 1950.

"*likely to satisfy few*": "He Made Rain, But—" (editorial), *Schenectady Gazette*, August 10, 1950.

"*The Commissioner now requests*": "Rain-Maker Gets Another Six Months," *New York Times*, August 18, 1950.

"*Certainly Dr. Howell*": "Six Per Cent Investment" (editorial), *New York Herald Tribune*, August 19, 1950.

18. Señor O'Dwyer

"*During the past month*": Harold H. Harris, "Mayor Named Mexico Envoy," *Brooklyn Eagle*, August 15, 1950.

"*the biggest police-bookmaker*": "Former New York Mayor, William O'Dwyer, Dies," *Cortland Standard*, November 25, 1964.

"*obsessed* with the investigation": "The Evolution of Mr. O'Dwyer from Mayor to Ambassador" (editorial), *Brooklyn Eagle*, August 16, 1950.

"*fair, fearless, and honest*": "Jury Kills Mayor's 'Witch Hunt' Charge," *Brooklyn Eagle*, August 16, 1950.

"It is hardly surprising": "Senor O'Dwyer in New Role" (editorial), *Utica Daily Press*, August 19, 1950.

"Yet there was a tinge": "Truman Taps O'Dwyer to Be Mexican Envoy," *Buffalo Courier-Express*, August 16, 1950.

"All of a sudden": Robert Ruark, "The O'Dwyer Departure: Jumped—or Pushed?," *Elmira Star-Gazette*, August 29, 1950.

"For example, should Mr. Impellitteri": Moscow, "Naming of O'Dwyer as Envoy."

"If, as now seems likely": Warren Moscow, "O'Dwyer Will Retire Aug. 31 to Go to Mexico as Envoy; City Votes on Mayor Nov. 7," *New York Times*, August 16, 1950.

several other possible hats: Harris, "Mayor Named Mexico Envoy."

"wide open race": Warren Moscow, "Flynn Starts Ticket Parleys, Calls Choices 'Wide Open,'" *New York Times*, August 17, 1950.

"He was an immigrant": "O'D, in Tears, Buries Hatchet with Cashmore," *Brooklyn Eagle*, August 17, 1950.

"The 'deal,' reportedly": "Sinnott Presents Boro Mayoralty Choices to Powwow," *Brooklyn Eagle*, August 23, 1950.

"at a not far distance": Violet Brown, "50,000 at Hall Hear O'Dwyer's Farewell," *Brooklyn Eagle*, August 31, 1950.

"The post to which": "Text of O'Dwyer's Farewell Address," *New York Times*, September 1, 1950.

"May you continue to progress": Brown, "50,000 at Hall."

"It was not, however": Paul Crowell, "Mayor Leaves City After Ceremonies Seen by Thousands," *New York Times*, September 1, 1950.

"I hereby resign": Paul Crowell, "Impellitteri Takes Full City Powers," *New York Times*, September 3, 1950.

"a slightly built": Robert D. McFadden, "Vince Impellitteri Is Dead; Mayor of New York in 1950s," *New York Times*, January 30, 1987.

"amiable but slow witted": Caro, *The Power Broker*, 788.

"well known in advance": Warren Moscow, "Pecora Nominated by 9,000 in Garden," *New York Times*, September 10, 1950.

"I certainly don't intend": James A. Hagerty, "Impellitteri to Run 'Independently' If Kept off the Democratic Ticket," *New York Times*, September 5, 1950.

"You would be surprised": "Impellitteri Backer Active in Times Sq.," *New York Times*, September 15, 1950.

"Impellitteri privately promised": Caro, *The Power Broker*, 789.

"This is a community": "Moses Will Support Impellitteri in Race," *New York Times*, October 25, 1950.

"appointed for life": "Impellitteri Sees Victory, Free Hand," *New York Times*, November 7, 1950.

"The Man Who Cannot": Impellitteri campaign ad, *Brooklyn Eagle*, October 26, 1950.

19. Autumn

"torrential downpour which flooded": "Weekend Downpour Floods Boro, Queens," *Brooklyn Eagle*, August 21, 1950.

"Sometimes they like it": "Rainmaking Splits State's Farmers," *New York Times*, August 23, 1950.

"The planes, not on": "Dr. Howell About to Open New Onslaught on Clouds," *New York Herald Tribune*, September 14, 1950.

"About 2 P.M. yesterday": "Rain-Making Test Is Balked by Rain," *New York Times*, September 14, 1950.

"He said that a few": "Carney Suspends Dry Days and Bans," *New York Times*, September 15, 1950.

"with the approach": "Water Use Drops on Last Dry Day," *New York Times*, September 16, 1950.

"until further notice": "Carney Suspends Dry Days," *New York Times*.

"Praising the public": "City Ends 'Dry Days' and Eases Water Use Rules," *Brooklyn Eagle*, September 15, 1950.

"Thus," Langmuir said: Laurence, "Seeded Rain Noted."

"has changed the face": Doug Burns, "Langmuir Tells Scientists Seeding of Clouds Changed Face of U.S. Weather Map," *Schenectady Gazette*, October 13, 1950.

"It is thought that some": Bliven, *Preview for Tomorrow*, 95.

"may pour additional trouble": "Political Problems Seen in Rain-Making," *New York Times*, November 18, 1950.

"major anomalies": Burns, "Langmuir Tells."

"The whole experiment": "Tomorrow's Weather," *Fortune*, May 1953.

"much vaunted as the answer": "Weather Mechanics" (editorial), *New London Day*, October 22, 1950.

"*Although Dr. Langmuir didn't say*": Earl Ubell, "Langmuir Hints Cloud Seeding Detours Rain," *New York Herald Tribune*, October 13, 1950.

"*All hope of an early end*": "Allied Armies Forced Back as 130,000 Reds Join Drive," *Brooklyn Eagle*, November 3, 1950.

"*The only hope for*": Harold H. Harris, "City, State Poll Outcomes Seen Pivoting on Boro Vote," *Brooklyn Eagle*, November 6, 1950.

"*precedent-shattering victory*": Leo Egan, "Impellitteri Elected Mayor by 219,527," *New York Times*, November 8, 1950.

"*The oath-taking required*": Charles G. Bennett, "Impellitteri Takes His Oath as Mayor on City Hall Steps," *New York Times*, November 15, 1950.

"*Mr. Carney hopes*": "City Water Use Tops Rainfall, Supply 63.4%," *New York Herald Tribune*, November 15, 1950.

20. Thanksgiving

"*The Thanksgiving Eve crash*": "L. I. Rail Road Wreck Toll Climbs to 77 Dead, 329 Hurt," *Brooklyn Eagle*, November 24, 1950.

"*the grimmest disaster*": "75 Known Dead in L. I. Wreck in Richmond Hill," *New York Times*, November 23, 1950.

"*War clouds in Korea*": "City's Thanksgiving Is Quiet Owing to War and Disaster," *New York Times*, November 24, 1950.

"*A few snow flurries*": "Frigid Blast Sweeping Toward City from West," *Brooklyn Eagle*, November 24, 1950.

"*PORT ARTHUR, Ont.*": "Lake Ships Shelter in Storm," *New York Times*, November 25, 1950.

"*As a result, a low pressure*": "Air Masses Meet, Start Big Storm," *New York Times*, November 26, 1950.

"*Instead, the disturbance*": "Storm Worst in 37 Yrs., Says Weather Bureau," *Schenectady Gazette*, November 27, 1950.

newspaper weather map: "Main Course of Storm and Area Affected," *New York Times*, November 26, 1950.

"*Tons of snow continued*": "Worst Storm in 37 Years Hits Ohio; Blizzard Halts Business, Industry," *New York Times*, November 26, 1950.

ONE SNOWSTORM THAT MISSED: "One Snowstorm That Missed Buffalo," *Buffalo Evening News*, November 25, 1950.

"freakish electrical storm": "Bolts Fire Power Plant," *New York Times*, November 26, 1950.

"beaten and bruised": "6 Dead as Storm Hits 90-Mile Pace," *Syracuse Post-Standard*, November 26, 1950.

"the most violent of its kind": "Atlantic Coast Hard Hit by Storm," *Schenectady Gazette*, November 27, 1950.

"Piers on the Brooklyn waterfront": "Gale Lashes Boro Wreaking Heavy Property Damage," *Brooklyn Eagle*, November 25, 1950.

"As a consequence trains": "Floods, Loose Wires Peril Autos—Transit Service Disrupted," *Brooklyn Eagle*, November 26, 1950.

"What was predicted": "Violence from the Skies" (editorial), *New York Times*, November 27, 1950.

"brackish water swirling": Mac R. Johnson, "Thousands Driven from Flooded Homes," *New York Herald Tribune*, November 26, 1950.

"By noon the city": "Gales and Rain Ravage City Area, Killing 23," *New York Times*, November 26, 1950.

"The trip, which normally": "Gale Leaves Trail of Shattered Glass, TV Aerials, Signs," *Brooklyn Eagle*, November 26, 1950.

"On the other hand": Ralph Chapman, "Storm Is Third of Kind to Hit City in 12 Years," *New York Herald Tribune*, November 26, 1950.

"There was a severe wind": "Record Flood Greatly Damages Mountain Area," *Catskill Mountain News*, December 1, 1950.

"The Kingston Police Department": "Kingston Escapes Main Fury of Storm Sweeping Nation," *Kingston Daily Freeman*, November 25, 1950.

"Esopus Creek went on a rampage": "Two Dams Burst, Cause Million in Damage in Shandaken; 226 Dead Along Seaboard," *Kingston Daily Freeman*, November 27, 1950.

"An even more striking gain": "Phoenicia Area Is Drenched with 4.63 Inches of Rainfall," *Kingston Daily Freeman*, November 27, 1950.

"Ulster county police said": "Atlantic Coast Hard Hit," *Schenectady Gazette*.

"It might be a coincidence": "What Say, Doc, Snow for Christmas?," *Kingston Daily Freeman*, November 22, 1950.

"What are they trying": "'Cloud-Seeding' Is Claimed over Area," *Kingston Daily Freeman*, November 27, 1950.

"It was explained yesterday": "City Seeded Cloud in Midst of Storm," *New York Times*, November 28, 1950.

FLOOD'S BEST JOKE: "Flood's Best Joke," *Catskill Mountain News*, December 1, 1950.

"extraordinary": Howell, "Precipitation Stimulation Project," 94–95.

"Apparently unabashed by": "New York Water Level Up to 76.4% of System Capacity," *Kingston Daily Freeman*, November 30, 1950.

"When will the rainmaker": Mrs. James Smith, "The Rainmaker's Flood Which Damaged Pine Hill," *Catskill Mountain News*, December 8, 1950.

"The ill wind and the rains": "The Reservoir Rise" (editorial), *New York Times*, December 5, 1950.

"In the amount of energy": "Weather Control Called 'Weapon,'" *New York Times*, December 10, 1950.

"Did the Belleayre": Mountain Dew, *Catskill Mountain News*, December 15, 1950.

21. Winter

"We are where we are": "Reservoirs High and Going Higher," *New York Times*, December 13, 1950.

"Politics, rather than any": "Mayor Will Seek Ouster of Carney," *New York Times*, December 19, 1950.

"Frank Sampson phoned me": Paul Crowell, "Carney Ousted from Water Post; Supported Pecora Against Mayor," *New York Times*, December 23, 1950.

"purely a political move": "Purge by Mayor Claims Carney as First Boro Victim," *Brooklyn Eagle*, December 22, 1950.

"labored long and hard": "Water over the Damned" (editorial), *New York Post*, December 26, 1950.

"for the good": Charles G. Bennett, "Paduano Gets Carney's Water Job; Firemen Aiding Pecora Disciplined," *New York Times*, December 27, 1950.

"Rain-making experiments over": Walter Lister Jr., "Rainmaker's Contract Expiring in Week; City Unlikely to Renew," *New York Herald Tribune*, February 14, 1951.

"the meteorologist is expected": "City's Rain Making Likely to End Soon," *New York Times*, February 14, 1951.

"carried out on a regional": Lister, "Rainmaker's Contract Expiring."

"$100-a-day-when-he-works": "Old Man Winter Isn't Dead Yet; Merc Skidding," *Brooklyn Eagle*, February 14, 1951.

"*We're sorry to see*": "Rainmaker Rained Out" (editorial), *New York Herald Tribune*, February 15, 1951.

"*to get more definitive*": "Rainmaker Wants to Continue Tests," *New York Herald Tribune*, February 19, 1951.

"*The results are promising*": Paul F. Ellis, "Rain-Maker Washed Up, Believes Work Helped Some," *Tonawonda Evening News*, February 21, 1951.

"*He went into some detail*": Robert D. Elliott, "Experience of the Private Sector," in Hess, *Weather and Climate Modification*, 71.

"*He urged Federal control*": "City Rain Maker Says He Lifted Fall 14%," *New York Times*, March 17, 1951.

"*about 17 percent*": Howell, "Precipitation Stimulation Project," 104.

"*has been suppressed*": "Water Use Sets 2-Year High in Face of Drought," *New York Herald Tribune*, July 18, 1951.

"*I haven't had an opportunity*": "Lawn, Street Sprinkling Curbed; Water Reserves Below '49 Level," *New York Herald Tribune*, June 29, 1951.

"*just a little lawyer*": "Herman Gottfried: 'Little Lawyer' with a Big Practice," *Binghamton Press*, June 20, 1961.

"*had no right to appear*": "Why Does NY Seek to Stop Gottfried?," *Catskill Mountain News*, January 4, 1957.

"*The claimants contend*": "City Sued on Rain-Making," *New York Times*, February 17, 1951.

"*It is believed that*": "Damage Claims Filed Because of November Flood," *Catskill Mountain News*, February 16, 1951.

"*interfering with atmospheric*": "Shandaken Blames Howell, Asks $167,500 for Damages," *Kingston Daily Freeman*, February 19, 1951.

"*The document held*": "Kingston Group Sues for Damage by Rains," *New York Times*, June 30, 1951.

"*no present or future intention*": "Thirsty for a Title?" (editorial), *Albany Knickerbocker News*, August 24, 1951.

"*The shoe was on*": "City Now Skeptic on Rain-Making; Damage Claims Total $2,138,510," *New York Times*, November 5, 1951.

"*to bring down rain*": "Rainmaking Abandoned" (editorial), *New York Times*, November 10, 1951.

"*and as incredulous clerks*": "Suits Pour on City for Its Rainmaking," *New York Times*, December 22, 1951.

in Australian pounds: "Rainmakers Face Huge Claims," *Sydney Morning Herald*, December 22, 1951.

"he would serve a note": "City Served in Suit over Rainmaking," *New York Herald Tribune*, December 22, 1951.

"The city's rainmaker": "Got a Gulley-Washer," *Cape Girardeau Southeast Missourian*, January 8, 1952.

"a mechanical error": "City Stems Deluge of Rainfall Suits," *New York Times*, February 19, 1952.

"a mere oversight": "Court Rules Against Suits for Damage in Rainmaking," *Schenectady Gazette*, February 19, 1952.

"form above substance": "NYC Can Be Sued for Rain Making in 1950," *Hancock Herald*, June 19, 1952.

"Photographs were offered": "Reserves Decision in Damage Case on 'Cloud Seeding,'" *Kingston Daily Freeman*, March 1, 1952.

"the plaintiff is unable": "Seek to Examine NY City Officials on Rain-Making," *Schenectady Gazette*, September 28, 1955.

"Some $2,500,000 in lawsuits": Charles R. Douglas, "$2,500,000 in Lawsuits Still Pending Decade After Seeding of Clouds in Area," *Kingston Daily Freeman*, March 19, 1960.

"where villages and hamlets": Kirk Semple, "On Flood Plain, Pondering Wisdom of Rebuilding Anew," *New York Times*, September 4, 2011.

"Farmers, business men": Charles Grutzner, "Upstate Water Claims Flood City," *New York Times*, July 28, 1957.

"a David bringing the Goliath": "Herman Gottfried: 'Little Lawyer,'" *Binghamton Press*.

Epilogue

"it would be a long time": Joy, "Rainmaker to Industry."

"Everything in America": Steinberg, "Cloudbusting in Fulton County," 210.

"devotes most of its attention": Jack Lotto, "Three Major American Firms Control Most of Multi-Million Dollar 'Rain Making' Business," *Lubbock Avalanche-Journal*, December 21, 1951.

"where a mining company": "Howell Says He Gave Rain to Peru," *Kingston Daily Freeman*, November 7, 1952.

"When our rain stimulating": Joy, "Rainmaker to Industry."

"Besides asking the public": "Stephen J. Carney, Led Water Bureau," *New York Times*, August 20, 1953.

"as things went along": "Wagner's Sweeping Victory Shows Great Voting Strength" (editorial), *Brooklyn Eagle*, September 16, 1953.

"In addition to his rain-making": "Dr. Irving Langmuir, 76, Dies at Cape Cod, Nobel Prize Winner," *Schenectady Gazette*, August 17, 1957.

"against legal proceedings": "Control Plan for Weather Washes Out," *Billboard*, June 6, 1953.

Howell Associates cloud-seeding operation: Orville, "Weather Made to Order?"

"who seems aptly named": Jack King, "Man Named Wetterer Has Trick Job of Rainmaking," *Greenfield Recorder-Gazette*, July 26, 1957.

"Cloud seeding is a top": Charles B. Douglas, "Weather Modification Study Asked for Area," *Kingston Daily Freeman*, May 24, 1961.

"Dutchess and Columbia County": "Placing Generators in Rainmaking Plan," *Kingston Daily Freeman*, August 6, 1962.

"While some areas received": "2-County Farmers Hire Rainmaker," *Albany Times-Union*, August 5, 1964.

He always recalled his stint: Stephen B. Howell, author interview, January 3, 2018.

"There can be no answer": National Research Council, *Scientific Problems of Weather Modification*, 3.

"This report was received": Elliott, "Experience of the Private Sector," in Hess, *Weather and Climate Modification*, 76.

So RAINMAKERS ARE PHONY: "So Rainmakers Are Phony After All," *Newark Star-Ledger*, November 22, 1964.

"Basic weather scientists": Wil Lepkowski, "Rainmakers Stir Up Storm of Confusion," *Syracuse Herald American*, June 27, 1965.

"Howell's seeding contracts": Kahan, "In Memoriam: Wallace E. Howell," 6.

"Wallace E. Howell of Howell": "Cloud Seeder Offers Water at Penny for 1,000 Gallons," *New York Times*, August 14, 1965.

"I'm interested just as": Homer Bigart, "Rainmaking Test Has Failed So Far," *New York Times*, August 21, 1965.

"worked to remove": Kahan, "In Memoriam: Wallace E. Howell," 6.

"Make Mud, Not War": Kabat, "Rainmaker's Flood," https://harpers.org/blog/2016/08/the-rainmakers-flood/.

"waging a systematic war": Seymour M. Hersh, "Weather as a Weapon of War," *New York Times*, July 9, 1972.

"It was disappointing": Kean, "Chemist Who Thought He Could Harness Hurricanes."

"idea of trying to throttle": Barnett, "Can We Engineer a Way to Stop a Hurricane?," https://news.nationalgeographic.com/2017/10/hurricane-geoengineering-climate-change-environment/.

"the chain of events": Matt Simon, "Could Scientists Use Silver Iodide to Make Snow for the Olympics?," *Wired*, February 19, 2018, www.wired.com/story/could-scientists-use-silver-iodide-to-make-snow-for-the-olympics/.

"keeps coming back": Andrew Moseman, "Does Cloud Seeding Work?," *Scientific American*, February 19, 2009.

"Evaluation methodologies vary": Naomi Oreskes and James Fleming, "An Ill Wind Blows Toward an Even More Inhospitable Climate," *Los Angeles Times*, October 30, 2003.

"In the core of the silver iodide": Pelley, "Does Cloud Seeding Really Work?," https://cen.acs.org/articles/94/i22/Does-cloud-seeding-really-work.html.

"flare trees that poke": David Biello, "Drought-Ridden L.A. Tries Rainmakers to Tap Storm Clouds," *Scientific American*, May 5, 2016.

"To get the complete picture": Scoles, "As Drought Looms," www.popsci.com/cloud-seeding-science.

"The rain dragons": Langewiesche, "Stealing Weather."

"The country's army": Central People's Government of the People's Republic of China, "China Leads World in Rainmaking," press release, June 4, 2006, www.gov.cn/english/2006-06/04/content_299936.htm.

"a procedure that made": Nick Macfie, "China's Artificially Induced Snow Closes 12 Highways," Reuters, February 19, 2009, www.reuters.com/article/us-china-snow/chinas-artificially-induced-snow-closes-12-highways-idUSTRE51I10X20090219.

"The NDRC approved": Josh Ye, "China Showers 1.15 Billion Yuan on Rainmaking Project for Parched Northwest," *South China Morning Post*, January 24, 2017.

"Although the saying goes": Kimbra Cutlip, "A Rain Maker's Passage," *Weatherwise*, September/October 1999.

Bibliography

Anderson, Clinton P., and Wallace E. Howell. "Should the U.S. Government Control the Rain Makers?" *Rotarian*, March 1951.

Arbuckle, Alex Q. "1906–1917: Building New York's Water Supply." *Retronaut*, May 7, 2016. https://mashable.com/2016/05/07 /building-new-york-water-supply/.

Barnett, Cynthia. "Can We Engineer a Way to Stop a Hurricane?" *National Geographic*, October 13, 2017. https://news.nationalgeographic .com/2017/10/hurricane-geoengineering-climate-change-environment/.

———. *Rain: A Natural and Cultural History*. New York: Crown, 2015.

Battan, Louis J. *Cloud Physics: A Popular Introduction to Applied Meteorology*. Mineola, NY: Dover, 2003. Reprint of *Cloud Physics and Cloud Seeding*. New York: Doubleday & Company, 1962.

Bliven, Bruce. *Preview for Tomorrow: The Unfinished Business of Science*. New York: Knopf, 1953.

Brimner, Larry Dane. *The Rain Wizard: The Amazing, Mysterious, True Life of Charles Mallory Hatfield*. Honesdale, PA: Calkins Creek, 2015.

Brooks, Stanley. "The Legal Aspects of Rainmaking." *California Law Review*, March 1949.

Buck, Rinker. "Drought City: The Real Scenario." *New York*, February 23, 1981.

Caro, Robert A. *The Power Broker: Robert Moses and the Fall of New York*. New York: Knopf, 1974.

Changing Times. "Every Man a Rain Maker? Artificially Induced Rain Is Definitely Here; It Has Impressive and Disquieting Possibilities." August 1950.

Chew, Joe. *Storms Above the Desert: Atmospheric Research in New Mexico, 1935–1985*. Albuquerque: University of New Mexico Press, 1987.

Christner, Brent C., et al. "Ubiquity of Biological Ice Nucleators in Snowfall." *Science* 319 (February 2008): 1214.

Clark, Edward J. "New York Control Curves." *Journal of the American Water Works Association* 42, no. 9 (September 1950): 823–827.

Coila, Bridget. "Changing the Weather." *Weatherwise*, May–June 2005.

Cummings, Arthur. "Helicopter Operations Within the New York Police Department." *AIAA Student Journal* 21 (Spring 1983): 38–40.

Drake, Lawrence. "Rainmakers Are All Wet: A Vigorously Negative Report on Man-Made Cloudbursts." *'48*, March 1948.

Earth Mover. "Driving the Shandaken Tunnel." December 1919.

Ellis, Edward Robb. *The Epic of New York City: A Narrative History.* Reprint ed. New York: Kodansha, 1997.

Fleming, James Rodger. "The Climate Engineers." *Wilson Quarterly*, June 2007.

———. *Fixing the Sky: The Checkered History of Weather and Climate Control.* New York: Columbia University Press, 2010.

———. "The Pathological History of Weather and Climate Modification: Three Cycles of Promise and Hype." *Historical Studies in the Physical and Biological Sciences*, September 2006.

Flinn, Alfred Douglas. "The World's Greatest Aqueduct: Water from the Catskill Mountains to the City of New York." *Century Magazine*, September 1909.

General Electric Company. "Dr. Vincent Schaefer Snow-Making Demonstration." Circa 1947. Posted on YouTube by Museum of Innovation and Science, Schenectady, NY, February 26, 2012. www.youtube.com /watch?v=2D5s2FlA_5k.

General Electric Review. "'Project Cirrus': The Story of Cloud Seeding." November 1952.

Groopman, Abraham. "New York Dependable Supplies." *Journal of the American Water Works Association* 42, no. 9 (September 1950): 827–828.

Harper, Kristine C. *Make It Rain: State Control of the Atmosphere in Twentieth-Century America.* Chicago: University of Chicago Press, 2017.

Havens, Barrington S., comp. *History of Project Cirrus.* G.E. report no. RL-756. Schenectady, NY: Research Publication Services, The Knolls, 1952.

Henderson, Thomas J. "Vincent J. Schaefer: A Remembrance," *Journal of Weather Modification* 26, no. 1 (April 1994): 163–165.

Hess, W. N., ed. *Weather and Climate Modification.* New York: John Wiley & Sons, 1974.

Hiltzik, Michael. *Big Science: Ernest Lawrence and the Invention That Launched the Military-Industrial Complex.* New York: Simon & Schuster, 2015.

Hischak, Thomas S. *100 Greatest American Plays.* Lanham, MD: Rowman & Littlefield, 2017.

Howell, Wallace. "Cloud Seeding and the Law in the Blue Ridge Area." *Bulletin of the American Meteorological Society* 46, no. 6 (June 1965): 328–332.

———. "The Growth of Cloud Drops in Uniformly Cooled Air." *Journal of Meteorology* 6, no. 2 (April 1949): 134–149.

———. "More Rain for New Yorkers?" *Weatherwise,* April 1950.

———. "Ours or Theirs?" *Water Power,* August 1954.

———. "The Precipitation Stimulation Project of New York City, 1950." *Journal of Weather Modification* 13, no. 1 (April 1981): 89–107.

Hughes, Patrick, and Stanley David Gedzelman. "Mysteries in the Clouds." *Weatherwise,* June 1995.

Irving, Washington. "The Catskill Mountains." *Littell's Living Age,* November 29, 1851.

Jones, Geoffrey. "Wild Goose Chase." *AIR Pictorial,* March 1997.

Journal of Weather Modification. "Bernard Vonnegut: 1914–1997." Vol. 29, no. 1 (April 1997): vi–vii.

Joy, Arthur F. "Rainmaker to Industry." *Flying,* September 1953.

Kabat, Jennifer. "The Rainmaker's Flood: The Quest to Control the Weather." *Browsings* (blog), August 31, 2016. https://harpers.org /blog/2016/08/the-rainmakers-flood/.

Kahan, Archie M. "In Memoriam: Wallace E. Howell, 1914–1999." *Journal of Weather Modification* 32, no. 1 (April 2000): 6–7.

Kean, Sam. "The Chemist Who Thought He Could Harness Hurricanes." *Atlantic,* September 5, 2017. www.theatlantic.com/science/archive/2017 /09/weather-wars-cloud-seeding/538392/.

Kellner, Tomas. "Kurt's Cradle: Kurt Vonnegut Was GE's PR Man Before Becoming a Bestselling Author." *GE Reports,* November 30, 2016. www.ge.com/reports/post/78009878650/kurts-cradle-kurt-vonnegut -was-ges-pr-man/.

———. "The Weather Men: How Kurt Vonnegut's Brother Tried to 'Abolish the Evil Effects' of Hurricanes with Science." *GE Reports,*

September 7, 2017. www.ge.com/reports/snow-men-cometh-kurt
-vonnegut-ice-nine-white-christmas-demand/.

Koeppel, Gerard T. *Water for Gotham: A History*. Princeton, NJ: Princeton
University Press, 2000.

Kurlansky, Mark. *Salt: A World History*. New York: Walker, 2012.

Lambright, W. Henry. "Government and Technological Innovation:
Weather Modification as a Case in Point." *Public Administration Review*
32, no. 1 (January–February 1972): 1–10.

Langewiesche, William. "Stealing Weather." *Vanity Fair*, May 2008.

Langmuir, Irving. *Atmospheric Phenomena*. Vol. 10 of *The Collected Works of
Irving Langmuir*, edited by Guy Suits. Oxford: Pergamon, 1961.

———. "Control of Precipitation from Cumulus Clouds by Various
Seeding Techniques." *Science* 112 (July 14, 1950): 35–41.

———. "Find Yourself This Summer." *Boys' Life*, June 1941.

Leonardo. "In Memoriam: Bernard Vonnegut." Vol. 31, no. 3 (1998): 208.

Liebling, A. J. *The Honest Rainmaker: The Life and Times of Colonel John
R. Stingo*. San Francisco: North Point, 1989.

Life. "Man-Made Snowstorm." December 30, 1946.

———. "Nick the Rain Maker." September 8, 1947.

———. "Solution to Water Shortage? Rain Makers' Success Shows How
New York Could Fill Its Reservoirs." February 20, 1950.

———. "U.S. Water: We Can Supplement Our Outgrown Sources—at a
Price." August 21, 1950.

Malone, Thomas F., ed. *Compendium of Meteorology*. Boston: American
Meteorological Society, 1951.

Marquardt, Meg. "October 13, 1947: A Disaster with Project Cirrus."
Earth, October 2010.

Meagher, Timothy J., and Ronald H. Bayor, eds. *The New York Irish*. Baltimore: Johns Hopkins University Press, 1997.

Morris, Jan. *Manhattan '45*. Baltimore: Johns Hopkins University Press, 1998.

Namias, Jerome. "Francis W. Reichelderfer: August 6, 1895–January 26,
1983." In vol. 6 of *Biographical Memoirs* by National Academy of
Sciences, 273–291. Washington, DC: National Academy Press, 1991.

National Research Council, Committee on Atmospheric Sciences. *Scientific
Problems of Weather Modification*. Washington, DC: National Academy
of Sciences–National Research Council, 1964.

National Research Council, Committee to Review the New York Watershed Strategy. *Watershed Management of Potable Water Supply: Assessing the New York City Strategy.* Washington, DC: National Academy Press, 2000.

Newell, Homer E. *A Recommended National Program in Weather Modification: A Report to the Interdepartmental Committee for Atmospheric Sciences.* Washington, DC: Federal Council for Science and Technology, 1966.

Novak, Matt. "Weather Control as a Cold War Weapon." *Smithsonian,* December 5, 2011.

Orton, Philip. "Hudson River or Estuary? (You May Be Surprised)." SeaAndSkyNY, March 23, 2011. https://seaandskyny.com/2011/03/23/is-hudson-river-an-estuary-or-a-river/.

Orville, H. T. "Weather Made to Order?" *Collier's,* May 28, 1954.

Patterson, Thomas W. "Hatfield the Rainmaker." *Journal of San Diego History* 16, no. 1 (Winter 1970): 2–27.

Pelley, Janet. "Does Cloud Seeding Really Work?" *Chemical & Engineering News,* May 30, 2016. https://cen.acs.org/articles/94/i22/Does-cloud-seeding-really-work.html.

Pendle, George. "The Rainmakers." *Financial Times Magazine,* June 15, 2012.

Petrosky, Henry. *The Road Taken: The History and Future of America's Infrastructure.* New York: Bloomsbury, 2016.

Plutarch. *Plutarch's Lives.* London: William Heinemann, 1920.

Popular Mechanics. "With the Sky Police." January 1932.

Powers, Edward. *War and the Weather, or, The Artificial Production of Rain.* Chicago: S. C. Griggs, 1871.

Protano, Louis. "William O'Dwyer, Mayor of New York City, 1945–1950: Immigrant, Mayor, Ambassador." Thesis, St. Francis College, 1975.

Qiu, Jane, and Daniel Cressey. "Taming the Sky." *Nature,* June 2008.

Rogers, Jedediah S. *Project Skywater.* Historic Reclamation Projects. US Bureau of Reclamation, 2009. www.usbr.gov/history/ProjectHistories/Project_Skywater_D1[1].pdf.

Rosenfield, Albert. *The Quintessence of Irving Langmuir.* Oxford: Pergamon, 1966.

Rothery, Agnes. *New York Today.* New York: Prentice-Hall, 1951.

Rubin, Victor. "Rain Made to Order." *Popular Mechanics,* March 1929.

Schaefer, Vincent J. "Irving Langmuir, Versatile Scientist." *Bulletin of the American Meteorological Society* 38 (1957): 483–484.

———. "The Production of Clouds Containing Supercooled Water Droplets or Ice Crystals Under Laboratory Conditions." *Bulletin of the American Meteorological Society* 29, no. 4 (April 1948): 175–182.

———. "The Production of Ice Crystals in a Cloud of Supercooled Water Droplets." *Science*, November 15, 1946.

———. *Serendipity in Science: Twenty Years at Langmuir University.* Edited by Don Rittner. Voorheesville, NY: Circle Square, 2013.

Schiermeier, Quirin. "'Rain-Making' Bacteria Found Around the World." *Nature*, February 29, 2008. www.nature.com/news/2008/080228/full /news.2008.632.html.

Scoles, Sarah. "As Drought Looms, Could This Team of Scientists Prove Cloud Seeding Works?" *Popular Science*, June 23, 2017. www.popsci .com/cloud-seeding-science.

Senate Committee on Commerce, Science, and Transportation. *Weather Modification: Programs, Problems, Policy, and Potential.* Washington, DC: Government Printing Office, 1978.

Shields, Charles J. *And So It Goes: Kurt Vonnegut; A Life.* New York: Henry Holt, 2011.

Silverman, Stephen M., and Raphael D. Silver. *The Catskills: Its History and How It Changed America.* New York: Knopf, 2015.

Soll, David. *Empire of Water: An Environmental and Political History of the New York City Water Supply.* Ithaca, NY: Cornell University Press, 2013.

Spencer, Steven M. "The Man Who Can Make It Rain." *Saturday Evening Post*, October 25, 1947.

Standiford, Les. *Water to the Angels: William Mulholland, His Monumental Aqueduct, and the Rise of Los Angeles.* New York: HarperCollins, 2015.

Stanford Law Review. "Who Owns the Clouds?" Vol. 1, no. 1 (November 1948): 43–63.

Steinbeck, John. *The Grapes of Wrath.* New York: Viking, 2014.

Steinberg, Theodore. "Cloudbusting in Fulton County: A Study on the Ownership of the Weather." *Michigan Quarterly Review* 32, no. 2 (Spring 1993): 209–230.

———. *Slide Mountain, or, The Folly of Owning Nature.* Berkeley: University of California Press, 1996.

Stradling, Stephen. *Making Mountains: New York City and the Catskills.* Seattle: University of Washington Press, 2007.

Strand, Ginger. *The Brothers Vonnegut: Science and Fiction in the House of Magic.* New York: Farrar, Straus and Giroux, 2015.

Taylor, Frank J. "They Make Rain." *Prairie Schooner,* Summer 1954.

Taylor, Hugh. "Irving Langmuir, 1881–1957." *Biographical Memoirs of Fellows of the Royal Society,* no. 4 (November 1958): 167–184.

Tillman, Barrett. *Forgotten Fifteenth: The Daring Airmen Who Crippled Hitler's War Machine.* Washington, DC: Regnery History, 2014.

Time. "Weather or Not?" August 28, 1950.

Titus, Robert, Lynn Woods, et al. *A Journey Through Esopus Creek.* Kingston, NY: Lower Esopus Watershed Partnership, 2011.

Townsend, Jeff. *Making Rain in America: A History.* Lubbock, TX: International Center for Arid and Semi-arid Studies, Texas Tech University, 1975.

Tuthill, Barbara. "Hatfield the Rainmaker." *Western Folklore* 13, no. 2 (April 1954): 107–112.

Tvedt, Terje, and Terje Oestigaard, eds. *The World of Water.* Vol. 3 in *A History of Water.* London: I. B. Tauris, 2006.

Van Burkalow, Anastasia. "The Geography of New York City's Water Supply: A Study of Interactions." *Geographical Review* 49, no. 3 (July 1959): 369–386.

Waldman, Jonathan. *Rust: The Longest War.* New York: Simon & Schuster, 2015.

Weatherwise. "Mysteries in the Clouds." June/July 1995.

Weickmann, Helmut, ed. *Physics of Precipitation: Proceedings of the Cloud Physics Conference, Woods Hole, Massachusetts, June 3–5, 1959.* Washington, DC: American Geophysical Union, 1960.

Wharton, Leo R. "Operation H_2O: And It All Started with a Well." *Bulletin* (General Contractors Association), March 1948.

Willoughby, H. E., D. P. Jorgensen, R. A. Black, and S. L. Rosenthal. "Project STORMFURY: A Scientific Chronicle, 1962–1983." *Bulletin of the American Meteorological Society* 66, no. 5 (May 1985): 505–514.

Wood, Graeme. "Riders on the Storm." *Atlantic Monthly,* October 2007.

Wyoming Water Development Commission. "The Wyoming Weather Modification Pilot Program Level II Study: Draft Executive Summary." December 2014.

Index

Williams, Thomas, 96
Winchell, Walter, 32
Winslow, Richard K., 50, 51, 55
Wolff's Lake Florence Cabins,
    Lakewood, NY, 88, 143

Wyoming Weather Modification
    Pilot Project (WWMPP),
    199

Yonkers, NY, 98